The Hydrangeas

By the same author

EFFECTIVE FLOWERING SHRUBS
THE MOUTAN OR TREE PEONY
THE FLOWERING SHRUB GARDEN
THE GAME OF BOULES (PETANQUE AND JEU PROVENCAL)

Hydrangea macrophylla 'Altona'

The Hydrangeas

Michael Haworth-Booth F.L.S.

Fifth, revised edition

Constable London

This edition first published by
Constable and Company Ltd
10 Orange Street London WC2H 7EG
Copyright © 1950, 1975 by Michael Haworth-Booth
First published 1950
New edition 1955
Third edition 1959
Fourth edition 1975
Fifth edition, completely revised 1984

ISBN 0 09 465760 2

Set in Linotron Bembo 11pt by
Rowland Phototypesetting Ltd
Bury St Edmunds, Suffolk
Printed in Great Britain by
St Edmundsbury Press
Bury St Edmunds, Suffolk

Preface to the Fifth Edition

As the fourth edition has sold out it seemed a good opportunity to revise the botanical classification to follow the findings of Miss Elizabeth McClintock, assistant curator of Botany of the California Academy of Sciences, San Francisco, in her monograph on the genus.

My objective, as stated in previous editions, is to stimulate the acquisition of some of the beautiful wild species of Hydrangea that have, so far, not reached civilization.

Careful examination of the herbarium specimens showed that the best forms of some of the most horticulturally valuable species had never been imported. For example, the form of *H. anomala* available here was a rather tender one from a low altitude with poor, scarce flowers; whereas E. H. Wilson's form from Hupeh had abundant and beautiful fragrant flowers. *H. scandens* was only represented in Europe by an ugly little variegated sport with poor, sparse flowers. The spectacularly beautiful *H. Oerstedii* from the Andes has never been imported at all, although it could obviously be the pride of every sunroom in the country.

Meanwhile the habitats of these lovely plants are increasingly destroyed, so that, unless something is done soon, all that will be left will be the dried and withered bits in herbariums.

Farall,   M.H.-B.
Roundhurst
Nr Haslemere, Surrey
February 1984

Contents

Illustrations

Colour

Black and white

Preface to the Third Edition

Since I first wrote this book in 1950, I have had another eight years of close contact with Hydrangeas, and on this page, available owing to a third edition, I have tried to condense the knowledge gained during that time.

The varieties Vibraye, Westfalen and, to a slightly less degree, Altona, have proved notably hardier and more free-flowering than all others as outdoor plants, doing well in all parts of the country wherever gardens are reasonably elevated above the worst early and late frost incidence.

It is evident that the modern flowering shrub bed can easily beat the herbaceous border for continuous flower display from April to September if interspersed Hydrangeas are used as to one quarter of the flowerable area, set off by *Genista æthnensis* and *G. virgata* with Hypericum Hidcote. These will amply colour the fourth flowering effect in July and August; the first effect in the period April and early May, the second in late May and early June, and the third at midsummer, being implemented by evergreen Azaleas; deciduous Azaleas; Rhododendrons, Roses, *Senecio laxifolius*, Philadelphus, Cistus and Genistas, in turn, successively, in acid soil; while in limy soils Chænomeles, early Helianthemums, Ceanothus, Tree Peonies and *Viburnum tomentosum* varieties take the place of the Azaleas.

New varieties of Hortensias have been tried out, but none has so far been found worthy to challenge the supremacy of those mentioned. Until a new break is obtained by proper hybridising it is indeed difficult to see how any progress in breeding can be expected. Where should we be with Rhododendrons if we had gone on merely breeding from the old clonal varieties?

Far the most fitting and favourable place for the Hortensias is against the foot of the walls of the dwelling house. Giving nearly five months of decorative and flowery effect, they are without a rival for this most important of all positions. Their best associates, Camellias, Evergreen Azaleas and Hypericum Hidcote, may be fittingly interspersed. In warm districts the south aspect, however, is too hot for Hortensias. *Ceanothus impressus*, Campsis Mme Galen, *Cistus lusitanicus decumbens, Senecio laxifolius, Raphiolepis umbellata, Camellia sasanqua* and Hebe Blue Gem or H. Autumn Glory, are good substitutes.

To sum up, I believe that the Hydrangea plays an essential part in the makeup of every good garden, properly associated with other flowering shrubs. Indeed, it is as important in late summer as the Rhododendrons, Azaleas, Tree Peonies and Roses are in their season.

1959   M.H.-B.

Preface to the Fourth Edition

As the years have passed, the Hydrangeas, among which I live, have told me a few more of their secrets but there has been a regrettable lack of new outdoor varieties apart from 'Preziosa', 'Diadem', 'Miranda', 'Blue Deckle' and one or two promising *paniculata* seedlings. There has been the usual crop of new names among the Hortensias but as no further hybridising has been done, but seed merely taken from the old cultivars, there does not seem to be any improvement on the best of the older varieties obtained over the last fifty years in the same way.

My conviction that *H. macrophylla* is a hybrid race is quite definitely proved, so far as I am concerned, *by a most convincing test*. A typical Lacecap variety – 'Whitewave' – was grown in a secluded greenhouse and carefully self-pollinated. A hundred seedlings were grown to flowering size and, as I expected, these show a variation far beyond what a true species could produce. There are pink, white and also blue-flowered Hortensias, pure *japonica* types with white flowers turning red in the sun; almost pure *maritima* types and large handsome Lacecaps rather like 'Mariesii'. But there is nothing quite like the peculiar 'Otaksa' with its orbicular, red-stalked leaves and intense blue, nodding flowers. It makes one think of Handel-Mazetti's tender, blue-flowered species from inland South China.

In the shrub-garden various forms of *H. heteromalla* and *H. paniculata* continue to sow themselves and the seedlings grow with great speed and vigour. Some have flowers better than those of the parents. On the other hand, *H. paniculata* var. 'Praecox' seedlings do not vary from the parent at all in either leaf or flower. *H. acuminata*, now often classified under *H. macrophylla* subsp. *serrata*, though referred to here by its old name, varies very slightly as to its seedlings, no more than one would expect from a pure species.

Owing to the kindness of Mr. Rokujo, I have got a plant of the wild species that I shall continue to call *H. maritima*. It is very tender and the flower-head has only a very few ray-flowers of dingy white; the young stems have no spots but are pure green. Like 'Joseph Banks' and 'Seafoam' it is noticeably more tolerant of limy soil than most Hortensias. Mr. Rokujo also sent me a nice rooted piece of the woodland species I have called *H. japonica*. This has tiny flowers like a miniature 'Grayswood', white turning red in the sun. The thin stems are red. It is quite hardy.

As regards the hardiness of the outdoor varieties, the advantage of the favourable microclimate at the foot of the walls of an occupied house continues to prove absolutely decisive in the colder parts of the British Isles and this also goes for air-drainage.

December, 1974   M.H.-B.

The puzzle, the resolve and the quest

Few flowering shrubs of any kind can approach *Hydrangea macrophylla* in popularity. In Britain, France, Germany, Holland and Belgium, hundreds of thousands of these plants are raised every year. To an only slightly less extent they are also raised in Switzerland, Italy, Denmark, Norway and Sweden. In North Africa and Egypt they are much grown and they are also among the glories of the gardens in the hills of India and in Ceylon and the East Indies. In China and Japan they are among the most widely grown of garden shrubs. Further south, they are pre-eminent in South African gardens and in Australia Hydrangeas are one of the most popular and effective of all flowering shrubs. In the Americas they are also to be found, where the climate allows, from one end of that great continent to the other. Thus it is strange that, until now, there has been no book in the English language dealing, as a specialist work, with the Hydrangeas.

When I first became interested in the Hydrangeas many years ago, I found their nomenclature in great confusion and muddle. It soon became evident that gardeners had had good reasons for giving up hope of getting them properly sorted out. Many fine Lacecap clonal varieties, commonly grown, carried half-a-dozen different names. Others had no names at all. Yet these Hydrangeas had not been bred in this country – each must have had a label of some kind when it arrived here.

The older varieties of the Hortensia type grown in gardens were so little distinguished that only about a couple of dozen of the newer sorts were commonly given their varietal names. The rest, though equally published originally, were lumped together as 'common Hydrangeas'.

After my first few years' work my prospects of sorting them out had actually dwindled. Every time that one variety got properly identified and classified more complications would arise in some other direction. Wise friends told me that it was really a hopeless task, but that it would be helpful if I could at least gather together and tabulate all the widely scattered previously written data on the Hydrangeas and present them in a nice handy little book at a reasonable price!

I began, then, to collect and tabulate these data. Unfortunately it soon became only too plain that such a book would be mere nonsense, for no two authorities agreed. In the end it was the plants themselves that came to my rescue. As they began to grow up around me their forms and foliages, the port and colour of their flowers, their reactions to their individual conditions of soil and aspect began to send to me a stream of messages, of visual signals, definite and true. They interested me until I was enslaved and I knew that I must devote myself to straightening out the tangle somehow, so that finally I could allot to each plant its proper name and position.

Thus resolved I began to ponder on the puzzles. Having digested what had been written and knowing that much of it, at least, must be wrong, I studied my plants the more attentively. I enlisted the help of Mr. H. K. Airy-Shaw of the Herbarium at Kew, of Mr. W. T. Stearn of the Lindley Library, of Mr. Jessep of Melbourne Botanic Gardens, of Mr. P. F. Morris, a botanist of that city, and of the officers of the Royal Horticultural Society. All most generously gave their help and counsel. Thunberg's original specimens were even procured by Mr. Airy-Shaw, by the courtesy of the authorities, from Uppsala and most carefully examined. Rare and ancient works were dug out from library shelves and, equally important, one or other of the team would keep me posted of any new matter on the subject that appeared.

My thanks are also due to the Arnold Arboretum and to the late Dr. Alfred Rehder for having kindly sent me an advance copy of his admirable Bibliography which was of great help in tracing the various citations.

In addition to observing my own collection, I ranged the country hunting Hydrangeas, toured the West and walked miles of nursery grounds, perhaps only to note the behaviour of some variety scorching on a windy plain among rows of conifers.

Gradually from dim beginnings the picture began to emerge. Sometimes the sole result of an exhausting day in a great library would be one tiny fact. Slowly the puzzles became more localised until they were merely blots upon a definite picture rather than large, confused areas. Thus we may now come to more precise details as to why it was not possible to accept and merely collate the previous writings.

A landmark in Hydrangea history was the beautiful *Flora Japonica* of Siebold and Zuccarini written in 1840. Little respected by botanists owing to its purely horticultural viewpoint, my gardener's mentality warmed to this work, so clear and forthright in its descriptions of the garden varieties. It seemed to me that here we had a fairly complete picture of the commoner garden types in cultivation in Japan at that remote date. Kinds not included would, I thought, probably be of later production. Later researches have strengthened this opinion.

The next most important source of knowledge was the scholarly note of E. H. Wilson on the Hortensias published in the *Journal of the Arnold Arboretum*, **4**: 233 [1923]. This great plantsman claimed that he had discovered at last the phylogenetic type (or original wild ancestor) of *Hydrangea macrophylla* in 1917, and in these notes he arranged and sorted out *H. macrophylla* and its varieties, and *H. serrata* and its varieties, in a masterly manner and with the full gamut of synonymy and references set out with invaluable clarity and completeness. But was he right in his distinctions between these species and their varieties? For long I dared not presume to question this most learned and gifted authority. But the more I studied the more certain I became that some of Wilson's theories were, at least, open to doubt; others I was quite unable to reconcile with conclusions reached after more detailed studies than he could afford in view of his wider interests.

Of 'Otaksa' which is the actual Japanese clonal garden variety of Thunberg's original specimens, Wilson said: 'this form . . . is one of the most popular with florists in America and Europe', whereas it is really better described as an ancestor, and he then proceeds to say that the variety 'Thomas Hogg' (an obvious hybrid with the woodland species in my opinion) is a form of that form.

H. macrophylla var. *rosea*, another important foundation clonal variety imported from Japan by Charles Maries in 1880, he renamed *H. m. f. Veitchii*, for no good reason that I can see. He then took this existing varietal name *rosea* which had been widely published and applied it to *H. japonica* Siebold. At the time the name *Veitchii* was properly and commonly in use for quite a different Hydrangea and thus further confusion arose.

Then came the most misleading of his new name combinations, one which led me astray for years. He took Siebold's *H. Belzonii* and also Hooker's *H. japonica cœrulea* and made them into one imaginary Hydrangea which he called a form of his newly discovered wild *H. m.* var. *normalis*. Hooker's plant is, of course, a very ancient Japanese clonal variety of patent hybrid parentage with *H. serrata*. Wilson admits that it is far hardier than other 'forms', as he will call them, of *H. macrophylla* and even stresses the superior hardiness of *H. serrata* but ignores the inference. *Belzonii* is actually another quite distinct hybrid clone notable for its leaves being produced in threes instead of in pairs.

But worse was to come. Regel in his *Gartenflora*, **15**: 290, t. 520 [1866], figured a pure *H. japonica* type, the variety *macrosepala*, with large white ray-flowers each with three deeply serrated sepals. Wilson firmly put this down as another 'form' of his wild *H. macrophylla* var. *normalis*, unknown until 1917. It was many years before I found a specimen of this ancient Japanese variety. It left no room for doubt. Wilson was surely wrong here; it has little or no affinity with the maritime species.

Again, he also made Siebold's *Hydrangea japonica* a form of his wild maritime species *H. m.* var. *normalis*. Surely, I thought, this cannot be right. Siebold told us that his *H. japonica* had been in cultivation in Japanese gardens since the most ancient times, centuries before *H. macrophylla* appeared. Even a glance at Siebold's illustration tells us at once that the plant has little or none of the maritime species in its composition either. Siebold points out that *H. japonica* is easily distinguished from 'Azisai' (a *H. macrophylla* var.) by its slender red stems and reddish leaves. In any event the term *rosea* proposed by Wilson should not have been used. It was in active usage at the time for Maries' Hortensia, with perfect validity as a varietal name.*

With regard to *H. serrata*, Wilson made one species of:

(1) A tiny wild Hydrangea with deep pink entire-sepalled, orbicular ray-flowers known as *H. Thunbergii*;

(2) A wild Hydrangea with pale pink or blue ray-flowers with four entire, even-sized sepals known as *H. acuminata*;

(3) A Hydrangea with the remarkable characteristic of serrated ray-flowers which always open pure white and then turn deep crimson whatever the soil, known as *H. japonica*.

Surely, I felt, if a species is to vary as widely as this, the term 'species' loses meaning. The extraordinary unvarying colour-change of the flowers of *H. japonica* under the action of sunlight, uniquely unaffected by the presence or absence of aluminium, seemed to me an absolutely clear-cut distinction.

Wilson said that the flowers of *H. serrata* are 'blue or white in flat or slightly convex corymbs 5–10 cm. across with 4 or 5 white, blue or pinkish petaloid, *orbicular, ovate* or *lanceolate* (my italics) rounded, obtuse or, more rarely, acute sepals'. This description would cover almost any known Hydrangea.

Describing *H. serrata f. rosalba*, Wilson said:

The sterile ray-flowers may be white or pink or in fading may change from one colour to the other. The petaloid sepals may be deeply or sparingly toothed or they may be entire. It is a variable and very common form and is often cultivated under the erroneous name of *H. japonica*. There are *also* [my italics] very many garden forms of this plant; indeed, most of them which have been raised in France and have dull green leaves have been derived from it; such as H. 'Impératrice Eugénie' (Carrière in *Revue Hort.*, 469 t. [1868]).

Was the name *H. japonica* really erroneous? I wondered. In any event I knew that there was no vagueness about the

* See p. 23.

flowers *possibly* changing from one colour to another. *Rosalba* is an ancient clonal garden variety which received the Award of Merit and, like other clones, does not vary at all unless it produces a definite branch-sport. Its ray-flowers always open white and turn crimson by the action of sunlight. Intermediate forms might have remained alive instead of gracefully disappearing so as to leave only the most differentiated types clearly distinguishable as species, but none the less, I thought, such vagueness as this is impractical from the horticultural viewpoint. The explorer, pushing past, might well think that these flowers were variable in colour; the gardener, living with them, knows their colour changes to be invariable.

Were the Hydrangeas of the woodland group sufficiently different in character to be labelled as specifically distinct, or would I have to put up with the use of *H. serrata* as an 'umbrella-name' covering three or more distinct types? If *Thunbergii* and *acuminata* were really wild plants could anyone believe that they could produce such an extraordinarily different flower type as *rosalba*? The Japanese have bred these particular Hydrangeas (not *H. macrophylla* which is chiefly a twentieth-century French product), since ancient history, but even so it did not ring true.

It seemed unlikely that I could ever get to the bottom of these mysteries unless I could get to Japan and study the wild plants which, Wilson says, are 'common throughout the length and breadth of Japan'. Not only common but grown in almost every garden, *yet apparently never pictured*. I spent many days in turning over page by page the most exquisite coloured Japanese drawings of plants, collected in vast albums, yet never a sight of *H. serrata* in any of its innumerable forms, if we accept the Wilson thesis, did I get in these.

It seemed improbable that Wilson's one species *H. serrata* could cover the original *H. serrata* which included the three specimens of Thunberg's *Viburnum serratum*, *H. Thunbergii*, and *H. acuminata*, which has pink-blue (depending on available aluminium) ray-flowers with four rounded, entire sepals, and also the very different type of *H. japonica*, which is Siebold's plant and akin to the varieties *rosalba*, *macrosepala*, 'Grayswood', *intermedia* and *pubescens* having three or rarely four much-serrated sepals that open white and turn crimson due to sunlight action and are unaffected by aluminium.

These three Hydrangea species, for such I deemed them to be, bequeathed their different characteristics to their progeny through many decades of crossbreeding, however entwined and tangled the skeins of the genes may have become. So that we even find the coupled characteristics of *H. japonica* – deckle-edged sepals and refusal to 'blue' – oftener than not in a hybrid that is the result of fifty permutations; but I am anticipating the next problem and giving away its answer.

Another point also required clearing up. Could I prove that my strong suspicion that the Hortensias were, in fact, of hybrid origin was indeed a certainty? E. H. Wilson stated that he had examined about fifty named varieties of the Hortensias and could find no trace of hybridity in any of them. Yet he had mentioned the obvious *H. serrata* type, 'Thomas Hogg', without comment and he must surely have seen 'Mme Emile Mouillère', and, if he had, I could not but doubt his whole investigation of that matter. To my gardener's eye, living with the plants, the Hortensias proclaimed their hybridity aloud. Great was my joy when, following a lead from Mr. P. F. Morris, of Melbourne, I sent a collection of mounted seed vessels to Mr. Airy-Shaw for his learned investigation and received his full agreement that in his opinion they were indeed hybrids beyond doubt. Later I received confirmation from the French pioneer breeders that settled the matter finally. The whole picture now suddenly began to assemble itself in an orderly manner at last.

It became evident that Wilson's whole arrangement rested upon a false premise. His wild maritime species could not be called a variety of *H. macrophylla*, for that name belongs to Thunberg's original, Japanese, hybrid, clonal garden variety 'Otaksa', and thus, also, *to all the mighty race of other hybrid Hortensias and Lacecaps*. They have long borne that name, wherever accuracy is valued, and thus, let us be thankful, no change of name becomes necessary there.

All along another puzzle had intrigued me. This was the

extraordinarily distinct character of Sir Joseph Banks' original clonal Chinese variety. The immense, massive growth, huge thick wedge-pointed, shiny leaves, the plant's extraordinary vigour in full exposure and its unique and patent enjoyment of seashore positions right down on the beach, where no other Hortensia would grow, pointed to something unusual in its ancestry. Flowering only from the terminal bud, and not from every side shoot as well like the woodlanders, Sir Joseph's plant was one of the pointers that helped me on my way. Here, surely was simply a globose-headed sport of Wilson's pure maritime species. On the coast the continental-bred Hortensias appeared as weedy little bushes beside the giant. Yet, inland, the former rioted, while 'Joseph Banks' flowered sparsely.

The Lacecap clone 'Seafoam', as I provisionally named it, puzzled me greatly. In appearance and behaviour it, alone, strongly resembled 'Joseph Banks' yet I knew that Wilson's maritime species had not been imported into this country. What was it and how did it get here? It was only as we were almost ready to go to press that this mystery was resolved. Mrs. Crawley, a friend in the Isle of Wight, sent me some flowers of 'Seafoam' saying that it appeared to be wild on the cliffs near Shanklin. Immediately we set off and there found many plants of this Hydrangea among vast specimens of 'Joseph Banks'. None, however, appeared to be indubitably feral plants. Evidently, in spite of showers of seed from plantations above the inviting earthy cliffs, the alkalinity of the soil prevented the germination of seedlings. A search for some hours in a number of localities was fruitless, but an ample reward was yet to come.

Hundreds of fine plants of 'Joseph Banks' dominated the littoral landscape and suddenly a misplaced flower of 'Seafoam' caught my eye, appearing from the middle of a somewhat unthrifty plant of 'Joseph Banks'. A careful examination revealed that this was a branch-sport, or, more explicitly, a reversion. 'Joseph Banks' was, indeed, a branch-sport of Wilson's pure maritime species which I shall later★ call *H.*

★ See p. 62.

maritima★, and now had sported back to its beginnings. At once our researches switched to the new angle, and by the end of the day three more cases were found in which this reversion had occurred, thus leaving no room for doubt.

In my inland garden I had cultivated *H. maritima* for over twelve years – as 'Seafoam' – and in that time I had seen only three flower-heads. Alongside my flowerless plants, × *H. macrophylla* revelled in foaming masses of closely packed corymbs. There, by the sea, *H. maritima*, like its variety 'Joseph Banks', was densely covered with flowers, though a plant of no great beauty except where the soil was rich and black and a tree gave some shade from the sun so much more frequently seen here during the month of July than in Japan.

H. maritima came to England, then, as a bunch of latent cells in the fabric of its massive globose-headed sport that reached these shores in 1789.

The necessary provision of a new name for the pure wild maritime species causes little bother to anyone. It is but little grown in Europe and botanical seemliness will surely be better served by *Hydrangea maritima* than by *H. macrophylla* var. *normalis*.

There were a few untidy little corners left in the story of the Hydrangeas, it is true. But the little facts mounted up until at length the picture became clear and, after all, there is no finality in the study of plants. If we make even a little progress it should be put on record, or we may die without even the beginning of our trail marked out as a help to those who may follow.

Almost the only point that still puzzles me is the fact that 'Otaksa' has orbicular leaves and, more significant still, so has the wild-looking Lacecap 'Azisai'. Is there, perhaps, a wild Hydrangea species still lurking unknown in the forests of Japan with orbicular leaves?† It would be delightful to go there to find out the answer to this and other minor problems,

---

★ Wilson calls it *H. macrophylla* var. *normalis*.

† Since writing, I have seen unnamed species, as Herbarium specimens, just as described, but from China.

impossible of solution without research in the native land of the garden Hydrangeas.

The sorting out of the three-hundred-odd Hortensia varieties was a difficult business. No complete list existed. Many old varieties were almost extinct and often no description other than the raiser's oft-recurring 'Corymbes énormes, beau rose vif' had apparently ever been printed. It was necessary to find living plants. In the course of these researches old varieties unknown to me 'in the flesh', so to speak, were discovered in old gardens, and then these had to be identified.

Now, there are probably several hundred learned persons in this country who can identify any species, but I doubt if there are half-a-dozen who can identify all the garden varieties. In the end, however, the names were traced for all but one or two of those found to be good garden plants and these exceptions are still under observation.

A variety has to be outstanding to persist year after year against the competition of the endless stream of new ones. Thus 'Joseph Banks', *rosea*, 'Domotoi', 'Mme Mouillère', 'Radiant' and 'Vibraye' must command our respect, for they have persisted where hundreds must have perished, but I think that there are even finer new varieties. The difficulty has been to test them all under outdoor conditions over a long enough period to make certain that the newcomers possess equal hardiness and vigour.

With the lapse of time and the horrors Europe has undergone, even the actual breeders could not always give me exact information as to their own older varieties. Somewhere among these, there probably lurks the perfect outdoor Hortensia; long discarded perhaps by the trade grower of pot-plants by reason of just those qualities that make it supreme for lusty outdoor growth.

It is my hope that, conning the descriptions of the varieties, travellers may bring back with them cuttings of intriguing sorts from abroad and try them in their gardens. If merely placed in an airtight preserving jar or polythene bag with a wisp of damp moss, the shoots travel quite safely. I have, indeed, even sent them to Australia with satisfactory results.

Having no soil attached and not being, as yet, plants, there should be no customs complications.

Let there be no mistake, outdoor Hydrangeas are the late summer glory of favourably situated gardens all over Britain west of a line from the far north-west of Scotland to Dover.* And they are the perfect substitute for bedding-out.

The most eminent private gardener of our time (and I must leave his identity to the reader's deduction) once asked how it was that I had become so much interested in the Hydrangeas. In part the answer, in foaming masses of purest blue, lay around us on that September day in his own incomparable garden, the perfect result of three generations of skilful planters. But the Hydrangea varieties did not include the best that a more exact knowledge now enables us to select. It was the importance of attaining, once and for all, order and clarity in a family of beautiful and invaluable plants for the late summer garden that impelled me to undertake the task. It was a singular piece of luck that this fascinating work lay undone, ready for me to do, and its accomplishment has been highly enjoyable.

In the chapter dealing with the species I have striven for botanical accuracy without the use of avoidable botanical terms. I have done this, not only because I am personally incapable of forming a mental picture of the plant from a botanical description, but because I have found, not without some satisfaction, that botanists also have the same difficulty. No horticultural work fully covering the Hydrangea species is known to me and it seemed to me to be time that this gap was filled. Many of the species which appear to be highly desirable for our gardens have never been secured, and I hope that my descriptions may induce plant-hunters to pay more attention to this genus in future. Some newly reported species are unavoidably omitted owing to lack of specimens or adequate information. ·

Thus, while far from complete, this work provides a hand-

* Even east of this line they grow well in innumerable favourably situated gardens, doing particularly well in Elgin and around the Moray Firth on the north-east coast. See also p. 199.

book embodying the results of many years' study in growing and sorting out these fascinating plants and of all the facts that I could find that had been previously written about them. It is hoped that one result may be that the varieties and hybrids of the Hydrangea family, having at last a 'studbook' of their own, may be henceforward accorded the respect given to a Rose or Rhododendron variety. I think that the individuality and charm of the best kinds will, at all events, be found to be equally pronounced. My aim has been to provide sufficient for the gardener and perhaps a jumping off point for the studies of the botanist.

Since my first edition of this work, a *Monograph of the Genus Hydrangea* has been written by Miss Elizabeth McClintock, Assistant Curator of Botany, California Academy of Sciences. The monograph is published by the Academy under their Proceedings, fourth Series, Vol. XXIX, No. 5, pp. 147–256, with 10 maps and 6 plates.

Briefly, her treatment of *Hydrangea* differs from previous ones in that only two sections, to be called *Hydrangea* and *Cornidia*, are recognised. The section *Calyptranthe* becomes a subsection. Then, *Hydrangea macrophylla* (which she holds is a pure species) is placed in a new subsection *Macrophyllae*, and *H. serrata* is described as a subspecies to include my *H. acuminata* and *H. Thunbergii*, while my *H. japonica* is attributed to *H. macrophylla* subsp. *macrophylla*.

As a Horticulturist one must bow to the dictum of the Botanist, and for the purposes of this book I have adopted Miss McClintock's classifications from her splendid 1957 *Monograph*, though I still express strong reservations about her classification of the three latter plants, as well as *H. maritima*. But a person familiar with the living plants which he grows must be expected to differ sometimes from those who have to judge the plants from brown, dried herbarium specimens. It would be difficult to find two Hydrangeas more distinct in habitat, growth, foliage and flower than the sea–coast species and the various woodland species. There are, of course, two lines of botanical thought and their followers have been described as the 'lumpers', who like to amalgamate many into one species, and the 'splitters', who like to retain species and even, when desirable, to create new ones. For the horticultural trade the 'lumpers' are ruinous. This is because the comparatively simple binomial has to have subspecies 'so-and-so' added and then, perhaps, a varietal name as well, and this, being too much botheration for nine customers out of ten, finally puts many a good plant out of commerce.

In the case of a genus which is purely of horticultural interest, lumping is particularly disastrous. For example, the excellent garden-worthy species *H. villosa*, *H. robusta* and *H. Sargentiana* have now to come under the horticulturally inferior species *aspera* as mere subspecies or forms, so that a customer would now have to ask for *Hydrangea aspera* forma *villosa*, etc. This will seldom happen and so another blow to good, exciting and beautiful gardening is struck.

For my part, I maintain my conviction that *H. macrophylla* is a hybrid race and am confident that posterity will one day confirm this view. After all, I am the Director of the garden growing the largest collection of hybrids and species of Hydrangea in Europe and am in daily contact with the living plants. This book deals purely with the decorative value of the Hydrangeas as components of the modern garden.

Having made this point, I now gladly pay just tribute to the enormous amount of skilled work in the best botanical tradition that has gone into the most valuable study of the Hydrangeas by Miss Elizabeth McClintock.

Chapter 1

The history of the Hydrangeas

The history of this plant goes back a very long way. Among the wooded hills and on the shores of Japan, among the beautiful Red or Black Pines leaning picturesquely away from the wind, among the fretted lava rocks or the myriad tiny islets often planted by nature with blood-red Azalea, smoke-blue Wisteria and the wild white Rose, like the scene of a fairy story, the ancestral wild Hydrangeas grow.

The Japanese are the only other race who, like the English, have a garden around every house, where there is room for one, as a matter of course. At some remote date, the Japanese took the Hydrangea into their gardens, and, revelling in their care, it one day produced a 'sport' or freak seedling. Instead of having merely a ring of large showy flowers around the little ones in the centre of the disc-shaped flower-head, it produced a ball-shaped head almost entirely composed of the big sterile flowers, such as appear only around the margin of the flower-head in the wild form.

The common Guelder Rose (*Viburnum opulus*) of our woods has a flower very like that of the wild Hydrangea and, just like it too, has produced a garden form with a ball-shaped flower-head. Now, in both cases the large ray-flowers that surround the tiny real flowers in the centre are not the true flowers that carry out the essential work of producing seeds and pollen. They are there simply for ornament, to attract the notice of the bees and other insects who transfer the pollen from one of the little true flowers to another, fertilising them so that they can produce seeds. In short, instead of having one kind of flower to do both jobs, the Hydrangea has two kinds in each inflorescence – one kind to advertise and another to produce. Thus, this garden form can only produce seeds from the very few tiny true flowers that one may find hidden among the big showy ones that have taken the material of their sexual organs to make, instead, larger sepals.

For many years the Hydrangeas bloomed in the gardens of the Japanese, unseen and unknown by anyone from the outside world. This was because when, in the sixteenth century, the Japanese learned of the degradation and destruction of the noble, but insufficiently mechanised, native races of the Americas and the islands of the Pacific consequent on the penetration of their countries by the white man, they closed Japan completely to foreigners.* It was undoubtedly a very wise course and, at all events, gardeners everywhere have cause to be thankful that they did so, for it preserved the Japanese gardens and the innumerable flowers, unsurpassed in beauty by any in the world, that they contained.

During this period the only intercourse allowed with the outside world was the Dutch trading post which the Japanese permitted to be established on an artificial island in Nagasaki harbour, known as Deshima. To this post came Carl Peter Thunberg, the learned Swedish botanist, in 1775. He was allowed to make one heavily chaperoned journey to Yedo (now Tokyo), but much of his information on the plant life of the country came from his careful examination of the flowers among the hay supplied from the mainland every week for the animals kept for the table on the islet.

Occasionally, however, Japanese friends would bring flowers for the foreign plantsman, and, one day in 1776, the strange flower we know as the Hortensia, or the garden globose-headed form of Hydrangea, arrived. Nothing quite like it was known so that it is not surprising that Thunberg, when he described it in his *Flora Japonica* (1784), named it *Viburnum macrophyllum* – the large-leaved Viburnum. At about the same time he received another Hydrangea. This was a more slender-branched, slighter plant whose wild prototypes, it was discovered later, instead of growing in open places on the

---

* The beginning was an edict of the Tokugawa Shoguns expelling foreign priests in 1587. By 1638 this became fully operative for all foreigners. Following on Commodore Perry's famous mission in 1854, effective re-opening took place in 1856.

seashore close to the waves, appeared among the hills and on the wooded slopes of the mountains all over Japan. This, Thunberg named *Viburnum serratum* – the serrated Viburnum. Through the kindness of the Kew Herbarium authorities I have been privileged to examine the actual specimens of both these plants which Thunberg originally sent to the herbarium of Uppsala. Photographs that I took of these may be seen in Illus. 1, 2 and 3. These specimens and the Latin description that Thunberg wrote in his little book *Flora Japonica* were soon almost forgotten and remained so until 1830. We will return to this later.

In the meantime, a Chinese garden variety of the Hydrangea was said to have been introduced into France. The botanist Jussieu published a work in 1789 in which he quoted another botanist, Commerson, as having named this plant Hortensia to commemorate a certain lady's name.★ There are four ladies who might have been involved – Hortense Barret, a young girl who accompanied Commerson on the Bougainville expedition; Madame Hortense Lepeaute, wife of a famous clockmaker; Madame Hortense de Nassau, daughter of the Prince of Nassau, and, finally, Queen Hortense herself. The matter has been argued over by learned men for many years and no finality has been reached. At about the same time another great figure in the botanical world, Lamarck, published a description of a plant in cultivation in Mauritius under the name of *Hortensia opuloides*.†

A living Hydrangea was at last brought to Kew, actually from China, by the agency of the great Sir Joseph Banks in the same year of 1789. Its strange green flowers, we are told, excited the curiosity of all – even in the Customs House. This Chinese clonal (that is to say, vegetatively propagated) variety is identical with the coarse, strong growing, rather tender, weak-coloured Hortensia with huge bun-shaped flower-heads and thick, shiny, bright green leaves, that is still to be found in great numbers on the coasts of Cornwall and the west and south of England today. To distinguish it from the hundreds

★ According to Jaume St Hilaire in *Nouveau Duhamel*, **3**: 98; 1805.
† *Encyclopédie Méthodique Botanique*, **3**: 136 [1789].

of other clonal garden varieties that we have nowadays, I call it 'Joseph Banks'. The name-history, however, does not end here. Sir James Smith, first president of the Linnean Society, then named Sir Joseph's plant *Hydrangea hortensis*.* In view of later events we should note that this name was illegitimate, not only because Thunberg had previously named what was, until now, thought to be the same plant, but because Smith, when publishing it, cited as synonyms *Primula mutabilis* Loureiro and (with a query) *Viburnum serratum* Thunberg.†

In 1829, von Siebold, part author of that superb and beautifully illustrated book, *Flora Japonica*, by Siebold and Zuccarini, described a Japanese garden plant thought to be the same, from the botanical viewpoint, as *Hydrangea hortensia*.

Now, all this naming and renaming was not valid according to the rules of Taxonomy. This word covers all that relates to the laws and principles of the classification of plants according to their natural relationships. They had all forgotten Thunberg and his specimens and, most important of all, his little book in Latin in which he first published a description of the plant.‡

The rules state clearly that the first properly published name given to a plant with a proper description is the one and only correct name, no matter what other botanists may have called it since. But the plant was not, after all, a Viburnum, it was a member of a new genus.

In 1830 Séringe, under the ægis of the great de Candolle, whose name all students should revere because he was the first professor to make his lectures, even on the most tedious subjects, so interesting that there was always a queue down the corridor instead of half empty benches, transferred Thunberg's *Viburnum macrophyllum* to the genus Hydrangea as *Hydrangea macrophylla*.§ In this way, most properly, he retained Thunberg's specific name which has, of course, priority over any later ones. Thus, at length, we arrive at the conclusion that the full correct name of the plant we are discussing, *if*

---

* *Icones Pictæ Plantarum Rariorum*, t. 12 [1792].
† *Flora Cochinchinensis*, 104 [1790].
‡ *Flora Japonica*, 125 [1784].
§ Séringe in De Candolle, *Prodromus*, **4**: 15 [1830].

the Chinese and Japanese clones are of the same species, is *Hydrangea macrophylla* (Thunberg) Séringe.

We must remember that these plants are garden varieties with a globose-head of mostly sterile flowers; one of the wild parent species at this time was still growing unnoticed and unknown on the shores of Japan. *Hydrangea serrata*, as the other alleged species mentioned now became, was also imported into Europe at about this date, the year 1830.

Now we will leave the history of the Chinese garden Hydrangea in England for a moment and recount what happened in France. At some date between 1796 and 1830 Thunberg's Japanese garden form of the Hydrangea was imported into France under the name of 'Otaksa' as well as the Chinese form. Now these two Hydrangeas, we should note, are quite different. The Japanese 'Otaksa' has rather slender shoots, small round leaves almost as broad as long, and a globular flower-head of rose pink, even-sized flowers with entire, even, rounded sepals. It makes a bush about 3 feet high only. The Chinese 'Joseph Banks', or *hortensis* Smith, has very thick, coarse, massive shoots, large elliptic, wedge-pointed, thick, very shiny leaves and an enormous bun-shaped head of flowers that open greenish-yellow and gradually turn to a pale pinkish tint. It soon makes a big bush 6 feet or more high and across. For the moment I won't go into the fact that in an acid soil both kinds will have blue flowers – one a bright Cambridge blue and the other a soft grey-blue. That is a matter which will be discussed later on. The point is that the two varieties had different qualities.

Actually, as we shall find later, 'Joseph Banks' is a globose-headed sport of the pure wild maritime Japanese species and Thunberg's plant is a hybrid of that species crossed with the woodland species.

Thus the nomenclature requires some changes. The Hortensias which, except 'Joseph Banks', are all hybrids, are × *H. macrophylla* (Thunb.) Séringe.* E. H. Wilson's wild prototype of the maritime species and its variety 'Joseph Banks' require legitimate names.

* The × before the name signifies that it is a hybrid, not a pure species.

1 Thunberg's original specimen of *Viburnum macrophyllum* (= × *Hydrangea macrophylla* var. 'Otaksa')

2 Thunberg's original specimens of *Viburnum serratum* (= × *Hydrangea serrata*)

3 Thunberg's original specimen of *Viburnum serratum* (= × *H.s. prolifera*)

In the section devoted to the wild species I propose for Wilson's wild maritime species the name of *Hydrangea maritima* Haw.-Booth and for the Chinese, globose-headed garden variety of this species the name *H. maritima* var. 'Joseph Banks'. I mention this point briefly at this stage because I shall henceforward refer to these plants by their new names.

Returning to our historical notes, we find that Carl Johann Maximowicz, a Russian in charge of the Czar's gardens at Petrograd, went to Japan, reaching Hakodate in September, 1860. Being allowed to botanise within 20 miles of the city by the Japanese authorities, he stayed there until the end of November, 1861. To Petrograd he sent back about 800 herbarium specimens, 250 kinds of seeds and many bulbs.

He reached Nagasaki on January 4th, 1862, and left on March 30th for Yokohama, where he stayed from April 4th to December 21st, 1862, when he returned to Nagasaki, where he remained until December, 1863. After a month's stay at Yokohama he returned to Europe *via* the Cape of Good Hope, arriving home in Petrograd on July 10th, 1864. He brought with him 72 chests of herbarium specimens, 300 kinds of seeds and also about 400 living plants, including many Hydrangeas, a most remarkable achievement in those times. Maximowicz himself visited the mountains of Hakone and employed an intelligent young Japanese, Tchonoski, to collect for him. In 1867 he published his *Revisio Hydrangearum Asiæ Orientalis* (*Mém. Acad. Sci. St-Pétersb.* sér 7, **10, 16**: 1–48). This was probably the first scientific monograph on the Hydrangeas. In it Maximowicz divided the species into groups of affinity. Like E. H. Wilson and others later, he considered that von Siebold had made too many species. This is possible as von Siebold was working with the garden forms rather than the wild plants, but undoubtedly their pruning left too few species to account for all the original wild types. Through Maximowicz's associate, Regel, many of the Hydrangeas that he brought back soon reached Europe.

Charles Maries, who was plant collecting in Japan and China for that great firm of nurserymen, Messrs. Veitch, of Exeter, sent back two Hydrangeas in 1879. One was a Japanese hybrid garden variety with a bun-shaped head of mixed true flowers of the fertile kind and larger, pale pink, sterile flowers mostly round the outside; it was named *Mariesii*. The other plant was named 'Rosea' and it proved to be the most important foundation variety in the production of the Hortensias later. 'Rosea' is a different clone from Thunberg's plant, whose varietal name is 'Otaksa'. The growth of 'Rosea' is taller and more slender, the inflorescence dome-shaped, the sepals longer, narrower and more pointed, the individual flowers varying in size. The flower colour is a still richer pink or, in acid soil, a brighter blue. The leaves are domed, oval and pointed, not rounded like those of 'Otaksa'. It shows a strong infusion of *H. acuminata* and is evidently a hybrid of multiple parentage. Monsieur A. Truffaut obtained this variety and took it to France where he exhibited it before the Société Nationale d'Horticulture in Paris in 1901. Later, in 1903, he showed it again, this time grown in acid soil with, consequently, flowers of a bright pure blue.

In England the plant had received scant notice, but in Europe it was at once acclaimed as a plant of the highest value. Both *Révue Horticole*, Paris (p. 544 [1904]) and *Gartenflora* (Dec. 1, 1904, p. 617) carried coloured plates and a descriptive article in the text. I have a specimen plant.

In 1924, E. H. Wilson, one of the greatest plantsmen of all time, who studied the Hydrangeas, unfortunately decided to rename this plant *H. macrophylla* var. 'Veitchii'. This caused much confusion because a flat-headed hybrid variety from Japan, with white flowers, was in widespread cultivation under that name, quite legitimately. For details of this plant see × *H. m.* 'Veitchii'.

About 1830, von Siebold and, later, certain friends and associates of Maximowicz and others, imported Japanese garden varieties of other Hydrangea species. These were of the woodland Hydrangea species whose three distinct extreme types sheltered under the name of *H. serrata*. One of the most valuable of these is described by von Siebold in his *Flora Japonica* as *Hydrangea acuminata* and is a superior form of a common species. It has a northern distribution in Japan, and, typically, has ray-flowers with four rounded, entire, equal-sized sepals which are pink in neutral or limy soil and blue in acid soil like those of *H. maritima*. The seed vessels are tea-urn shaped, being long and tapered to the base.

Another Hydrangea was described, actually a garden variety, 'Beni Kaku', as *H. japonica* by von Siebold. Typically, this form has large ray-flowers with three sepals whose margins are markedly serrated in a decorative manner and have the most distinct and extraordinary flower colouring. The ray-flowers open white and then, *absolutely irrespective of whether the plant is growing in neutral-to-limy or acid soil*, turn crimson *owing to the effect of sunlight*. They do not turn blue however much aluminium is available. The seed vessels are cup-shaped, almost round, and the leaves have large, shallow teeth. Both these Hydrangeas have the valuable quality of flowering freely from side shoots as well as from the terminal bud. Their stems are slender and their leaves matt surfaced and thin, often puckered and with prominent veins. Both grow very weakly in warm, sunny positions, *acuminata* turning its dwarfed leaves to a purple colour and *japonica* soon failing if much exposed.

A third species, *H. Thunbergii*, is much earlier flowering and much more dwarf in growth with exceptionally numerous, tiny, orbicular ray-flowers of unusually vivid pink and reddish shoots.

I shall say more about these species later, but the point here is that the characteristics of freer flowering, deckle-edged sepals, and refusal to blue, or early flowering and orbicular flowers of deep colouring which we shall find in the Hortensias bred later, came from an infusion of the 'blood' of these woodland Hydrangeas.

Thunberg's 'Otaksa' had been cultivated in Japan for quite a long time, but *H. acuminata*, *H. japonica* and *H. Thunbergii* and their hybrids had been favourite garden plants of the Japanese since the most ancient times. So long ago, in fact, that I thought that the pure wild prototype of *H. japonica* might have become extinct centuries since, only partly hybridised forms remaining. For my part, I have strong reservations about classifying plants with such distinct flower characteristics, so potent in heredity, as of the same species.

In a horticultural monograph, as opposed to a botanical one, I may, however, take advantage of the slight degree of botanical laxity permissible in a mere gardener to describe, without further ado, as species the three Hydrangeas which E. H. Wilson lumped together under the name of *H. serrata*.

I feel that I must take this course because Wilson's pronouncement – equally wrong in my view – that *Rhododendron obtusum* is conspecific with *Rhododendron Kæmpferi*, caused equally disastrous results to those other supremely beautiful and effective garden flowering shrubs, the evergreen Azaleas. These lovely plants, like the Hydrangea the result of centuries of selective breeding *for garden decoration* by the Japanese, so long the master gardeners of the world,★ have remained in unnecessary muddle and a mystery ever since. Let there be no mistake; it is largely due to Wilson's genius that we have the plants at all. But, as I have said before, there is really no finality in the study of plants. Progress is continually made and new facts come to light. We should have no hesitation in deciding to realign these facts into their proper relationships, when truth points the way.

E. H. Wilson stated that he saw no trace of hybridity in Hortensia varieties which he examined, but he was studying hundreds of other genera at the time and I have concentrated the work of many years on a few genera. My botanist friends, Mr. H. K. Airy-Shaw of the Herbarium at Kew who has helped me in many long and exhausting hours of research with the resources of that great institution, and Mr. P. F. Morris in Australia, that land where the Hydrangeas luxuriate as in Japan, are in complete agreement that the garden Hydrangeas are, indeed, hybrids between *H. maritima* and the woodland species. A more detailed study of the distinct types of the seed vessels has enabled them to assure themselves on this point, and, indeed, the great French pioneer breeders have since confirmed the fact beyond doubt.

To the gardener, the Hortensias proclaim absolutely clearly

★ In the seventeenth century they published lists of hundreds of varieties.

the genes of this or that prototype in their make-up. The German group of 'Deutschland', with their large-toothed leaves, deckle-edged sepals and reluctance to accept the blue, hark back to *H. japonica*. 'Mme E. Mouillère' with her white colouring turning pink, and toothed sepals, also shows her *H. japonica* streak. 'President Doumer' and 'Westfalen', with their shapely little rounded flowers of intense colouring and dwarf growth, speak of *H. Thunbergii*. 'Vibraye', with its hardy nature, free flowering from the side shoots, tall slender stems and rich and eager blue reminds us of *H. acuminata*.

H. acuminata and *H. japonica* garden varieties were first imported into England, it seems, by Messrs. H. Low, of Clapton Nurseries, in 1844, and also in 1846 by Messrs. Henderson, of Pineapple Nursery, Edgware Road, and by Mr. Knight, of King's Road, Chelsea. *H. Thunbergii* was introduced about 1869 and first flowered by Messrs. Cripps of Tunbridge Wells.

Thus the Hydrangeas available at the close of the nineteenth century were as follows: *H. maritima* var. 'Joseph Banks', × *H. macrophylla* vars. 'Rosea', 'Mariesii' and 'Otaksa', and sundry Japanese garden hybrid varieties and forms of the three wood-land species. Some of the latter were enumerated in *The Garden*, by W. J. Bean (**54:** 390, t. 1196, Nov., 1898), as follows:

★'Aigaku', fls. light blue, new Japanese variety.
'Ajisai', rosy blue to light blue, new Japanese variety.
'Benigaku', rose coloured, new Japanese variety.
★'Shirogaku', white, blue centre, new Japanese variety.
'Thomas Hogg', pure white, globose head, new Japanese variety.
Lindleyi (syn. *japonica roseo-alba*), new Japanese variety.
Acuminata (syn. *H. Buergeri*).
Stellata (with several sub-varieties).

I have been able to trace no illustration or more detailed description of the varieties starred so cannot identify these exactly, but 'Aigaku' may be *H. acuminata* 'Bluebird', and 'Shirogaku' the Hydrangea we know as 'Lanarth White'. Guesswork, however, is not good enough.

In addition, Japanese nurserymen had sent over one of the

several attractive garden varieties that they had of *Hydrangea involucrata*. Thus the stage is set for breeding to commence.

In 1903 those great nurserymen and plant breeders, Messieurs Lemoine, of Nancy, father and son, noticed that the tea-urn-shaped seed capsules of a flower of × *H. macrophylla* var. 'Mariesii' were full of ripe seeds and so they sowed them. The seedlings produced three very valuable Hydrangeas for outdoor culture as flowering shrubs. They brought these out in 1904. They were (1) a strong hardy plant with rose-pink (or blue in acid soil) ray-flowers which they named *Hydrangea Mariesi perfecta*, (2) a glorious tall variety with huge white serrated-sepalled ray-flowers which they named *H. Mariesi grandiflora*, and (3) a plant with smaller but even more markedly serrated-sepalled pink ray-flowers, apt to turn bluish at the slightest hint of acidity, which they named *H. Mariesi lilacina*. All had flowers with the flat, Lacecap type, flower-head composed of a central area of the small fertile flowers with a marginal ring of showy sterile ray-flowers of large size on the outside.

Now in those days it was thought that *Mariesii* was a distinct species and not just a garden variety of × *Hydrangea macrophylla*, or Messrs. Lemoine would never have broken the rules by calling a new variety by the same name, with trimmings, as the parent variety. What a lot of confusion this mistake caused we shall learn later. In fact, of course, the extra word tacked on at the end soon got lost or forgotten and so it was not long before there were four quite different clonal varieties going about, all called *Mariesii*. This shows us how sensible the rules governing plant naming are and that it is to everyone's advantage to keep to them. But to return to these Hydrangeas, when I at last elucidated the mystery and traced their origin I had to give each of them new names in accordance with the rules and we shall find them described later as × *H. m.* 'Bluewave', × *H. m.* 'Whitewave' and × *H. m.* 'Lilacina'.

Since that time, except for 'Beauté Vendômoise', no one, except the writer, has brought out a new Hydrangea of this type with a flat flower-head – the type we call the Lacecap to distinguish it from the mopheaded type which we call the

Hortensia. These two common terms have not, of course, real botanical standing. They are simply garden names which, in one simple word, tell us the shape of the flower at once.

E. H. Wilson, in his notes on the Hydrangeas published in the *Journal of the Arnold Arboretum*, **4**: 234 [1923], unfortunately decided to tack on the words var. *normalis* to all the garden-bred Lacecap clonal varieties then adding the name of the variety as a 'form'. Thus if we followed this inconvenient system we should have to call a garden variety bred by Messrs. Lemoine, × *Hydrangea macrophylla* var. *normalis* forma 'Bluewave'. As var. *normalis* was used as the name of Wilson's wild maritime species, which is only one of the parents of these hybrids, there can be no question of following him in this today.

Actually these earlier varieties had a story to tell us. They showed us the hybrid parentage of × *H. macrophylla*. In 'Bluewave' the genes of the maritime species were predominantly given birth, in 'Whitewave' those of *H. japonica*, while *lilacina* showed almost equally the characteristics of both those wild species with a dash of *H. acuminata* as well.

I feel certain that had Wilson known these varieties he would have grasped the hybridity of the Hortensias at once.

In about 1905, Messrs. Lemoine, having observed that, amid the mass of sterile flowers composing the Hortensia flower-head, a few true fertile flowers were to be found, sowed the seeds formed by these. In 1907 and 1908 they astonished and delighted the horticultural world with the fine new varieties 'Avalanche', 'La Lorraine', 'Mont-rose' (showing strong *H. acuminata* blood), 'Bouquet-rose' and 'Radiant', which were the first produce of their work. 'Radiant' is still a prized variety, with good reason.

A little later Messrs. Mouillère of Vendôme produced (by crossing 'Whitewave' and *rosea*) 'Mme E. Mouillère', and also, by a different cross, 'Souvenir de Mme E. Chautard' and 'Générale Vicomtesse de Vibraye', and some of these varieties are still among the finest for outdoor planting today, owing to the free-flowering nature inherited more directly from their woodland hybrid parentage.

After the First World War, M. Henri Cayeux, of Le Havre, and his son and successor, M. Louis Cayeux, came into the field and 'Merveille' was one of the earliest and finest products of their skill. Later came a spate of valuable new varieties produced by many breeders. The Hortensia became a common conservatory plant and was widely used as a pot-plant for house decoration but, like the Rose, the Camellia, the Viburnum and many other shrubs from warmer climates, its hardiness and effectiveness for outdoor growing as a garden flowering shrub in the south and along the seaboard were only quite recently discovered.

In Cornwall and the west country, on the other hand, Hydrangeas were planted in quantity in the open and thousands of plants of the old varieties such as 'Joseph Banks', 'Bluewave', etc., are still to be seen there in old gardens. Curiously enough, however, so far as my personal researches go, there are comparatively few plants of the rich-coloured modern varieties to be seen in the west. The reason is, no doubt, that hybrid Hydrangeas with woodland 'blood' do not stand up to the sea-winds like *H. maritima*.

Now the reader may well be puzzled because in all the foregoing history I have said nothing about the discovery of one of the wild prototype species from which the garden Hydrangeas were derived. All this time this wild ancestor was growing in thousands on the coast of Japan and apparently nobody published news of it until E. H. Wilson came along in 1917, and, having long been on the lookout for it, at once declared it as the original wild prototype of the garden Hydrangeas. He described the plant as growing on the very shore, sometimes among lava rocks above, but *always* under the full influence of the sea. The flowers, he says, were flat-headed, the sepals of the ray-flowers pink, blue or, more rarely, white.

Wilson sent seed and herbarium material to the Arnold Arboretum (Boston, Massachusetts) but, so far as I know, this truly wild plant was never sent to England as such. Like many of the hybrid Hydrangeas, grouped under × *H. serrata*, more than a hundred varieties of the beautiful June-flowering Azalea

– *Rhododendron indicum*, Wistarias, Cherries, flowering Plums, many varieties of Iris Kæmpferi and innumerable other exciting plants for our gardens, the better varieties of this race are still waiting in Japan for someone to go and bring them back alive. Bold and brave plant hunters plunge into the depths of unknown forests, scale perilous mountains or penetrate forbidden districts where there is a price on their heads, every year. But to go and bring back some of the fruits of a great ornamental gardening tradition, that, alone, is older than our own by several centuries, no, that they do not do. One day, perhaps, somebody will go and get these Japanese plants for us, but it has not been thoroughly done yet. The backers of expeditions are usually more interested in new species than new garden plants, even if these are already brought to perfection by four centuries of skilful propagation and selection, but I think that many Japanese garden plants still unknown here would prove more valuable than any wildlings.

Many authorities state, somewhat vaguely, that *H. macrophylla* (presumably meaning Wilson's *H. m.* var. *normalis* which we now call *H. maritima*) is a native of China as well as of Japan, but Alfred Rehder of the Arnold Arboretum, who was in a better position to know all the facts than anyone else, informed us in *Plantæ Wilsonianæ** that he knew of no case in which this plant was reported as genuinely wild in any part of that country. But there are, I think, unnamed species very close to *H. japonica*.

It must be just a coincidence that the Chinese garden variety of Hortensia, *H. maritima* var. 'Joseph Banks' has the 'Chinese look': that coarse, strong, faintly toadlike yet strangely attractive flavour that characterises many lovely plants from China such as *Campsis grandiflora*, *Magnolia denudata* and *Chænomeles speciosa*. It is very distinct from the more delicate and refined grace of such typically Japanese plants as Hydrangea 'Grayswood', *Prunus serrulata* 'Yoshino' or Azalea 'Hinomayo'. Strangely enough, there is an almost exact parallel in the case of the Chinese double, white-flowered, garden variety of the

* *Plantæ Wilsonianæ* edited by Charles Sprague Sargent, **1**: 38 [1913].

Japanese Cherry, unknown in Japan and with a pronounced 'Chinese look'.

There is little of the 'Chinese look' about *H. maritima* itself, even seen issuing as a reversion out of a plant of 'Joseph Banks'. Perhaps the Chinese bred it for some years before it sported into opulence. Perhaps a passing junk spotted the great flower full blown on the shore among the spidery heads of the ordinary kind. Probably we shall never know.

The only Hortensia-breeding, in England, that I can trace was one sowing by the late H. J. Jones, of Hither Green, about the year 1925. He exhibited many varieties raised on the Continent at the Royal Horticultural Society's shows, in particular outstanding displays at the Chelsea Show in 1924 and 1925 which gained well-deserved Gold Medals. He showed his own seedlings in thumb-pots in 1926 and in 1927 many received awards. These are listed in the section devoted to the varieties. The nursery suffered much during the late war and became derelict after the deaths of the proprietor and his widow. Although good average sorts, only a few of Jones's Hydrangeas now survive in commerce. Examples of these are 'Miss Phyllis Cato', 'Mrs. R. F. Felton' and 'Mrs. A. Simmonds'.

I believe that the history of the Hydrangea, as an outdoor flowering shrub for the garden, is only just in its beginnings. Leading French breeders have told me that, among their annual batches of seedlings, Hydrangeas with a flat, Lacecap flower-head of deep rich colouring have appeared. As their target was still larger heads of the globose type, these seedlings were discarded. It was thought that they would not appeal to the florist. In the old days this was probably true enough, but nowadays we have many florists who have brought to their craft an original artistry and good taste which has caused quite a revolution. They have been quick to appreciate the possibilities for decorative use of the beautiful 'drawing' of the heads of the Lacecaps, with their exquisitely shaped, bold ray-flowers.

The Hortensia, unrivalled as a pot-plant for decorative use, will, also, I think, be vastly improved. An outcross is really overdue. Some of the allied wild species, discovered long after the early breeders began, have far more freely borne, larger, or more vividly coloured flowers than the ancestors of the Hortensias.

These matters and the whole question of breeding are dealt with in the chapter entitled 'The Breeding of Hydrangeas'.

In compiling the history of the Hydrangeas I have carefully studied the following works: *The Hydrangeas*, by E. H. Wilson; *Journal of the Arnold Arboretum*, Vol. 4, pp. 233–235, 1923; *Hydrangea et Hortensia*, by Marcel Ebel, Baillière, Paris, 2nd Edition, 1948; *Manual of Cultivated Trees and Shrubs*, by Alfred Rehder, Macmillan, New York, 2nd Edition, 1947; the Journal of the Royal Horticultural Society from 1904 to 1949; *Flora Japonica*, by P. F. Siebold and J. G. Zuccarini; *Trees and Shrubs Hardy in the British Isles*, by W. J. Bean; a *Monograph of the Genus Hydrangea*, by Elizabeth McClintock, San Francisco, 1957; and a number of other works mentioned in the text later. These authorities hold widely different views and my conclusions are arrived at after perusal of the works cited and a personal study of the plants themselves extending over many years.

Chapter 2

The species of Hydrangea

The name Hydrangea comes from the Greek *hydro*, water and *angeion*, a vessel,* alluding to the shape of the seed capsule of the first species discovered. In all, there are about thirty-five kinds in cultivation and the genus is generally included in the Saxifrage family, like the Carpentarias, Philadelphus, Deutzias, Escallonias and Ribes, though some authorities now make a new family of Hydrangeaceæ which also includes Philadelphus and Deutzia.

Roughly speaking the known wild Hydrangeas comprise six species in North America, twenty-seven in South America (all members of the Cornidia section), thirty in the Himalayas, China and Korea, twenty-two in Japan and Formosa, six in the Philippines and an unknown number in Java and the East Indies. A few range right across country from Japan to the Himalayas.

An important point to bear in mind regarding these hydrangeas is the immense variation in garden value of the various forms of each species. Many of these offer every gradation between worthless forms with insignificant flowers and very beautiful ones, whose inflorescences are large and exquisitely shaped and tinted. Unfortunately the forms in cultivation here are by no means always the best and examination of a wide selection of herbarium material makes this very plain.

In the Kew and British Museum Herbariums there are many specimens of Hydrangea species, so far unnamed, sent back from China (from Mo-kan-shan, Chekiang, in particular) and some of these appear to indicate that forms distinct from, but very close to *H. japonica*, *H. acuminata* and *H. Thunbergii* occur

* Thus, the middle syllable of the names should be pronounced 'ran', not 'rain', which makes an ugly-sounding word out of a pleasant one.

in that country. (These are ascribed in Miss McClintock's new classification to *H. macrophylla* subsp. *macrophylla*; the latter two to *H. macrophylla* subsp. *serrata*.) Little is known about the Formosan species beyond the brief notes given, and I have omitted most Javanese species owing to lack of information and the fact that they would probably be too tender for cultivation in this country.

The Hydrangeas are divided into two main sections: Hydrangea and Cornidia. Hydrangea contains the deciduous, shrubby species, and includes Calyptranthe, the deciduous climbing *H. petiolaris* and *H. anomala*; Cornidia the evergreen climbers and trees.

The sections are divided into subsections and some of these are further divided into series.

Following in part, Maximowicz's arrangement (*Mém. Acad. Sci.*, sér. **7, 10, 16:** 6 (*Rev. Hydrang. As. Or.* [1867]), Adolf Engler's in *Die Natürlichen Pflanzenfamilien* (2nd edition, **16a:** 207, etc. [1930]) and also Alfred Redher's setting out in *Plantæ Wilsonianæ* and his *Manual of Cultivated Trees and Shrubs*, as these best apply, and Elizabeth McClintock's *Monograph of the Genus Hydrangea* (Proc. Cal. Acad. of Sci.), 4th series, vol. xxix, no. 5, 1957, we may arrange the species as set out in the following pages. To avoid misunderstanding, the authorities for the names of each species are included, in accordance with botanical practice and, as there is a very strong family resemblance in their members, the subsections are given prominence.

Section I HYDRANGEA

Subsection 1 AMERICANÆ

Inflorescence mostly flat, rarely paniculate. Fertile flowers white. Petals of fertile flowers deciduous or persistent. Petals of sterile flowers white, turning reddish. Ovary inferior. Fruit partly hemispherical, truncate at apex. Seeds wingless.

This is the subsection of which *H. arborescens* is probably the best known.

4 *Hydrangea arborescens grandiflora*

Hydrangea arborescens Linnæus is an upright, bushy, but straggling, grower usually about 3 or 4 feet high. The name is misleading as many more recently discovered species are much more arborescent, or tree-like in character. The leaves are ovate (that is to say, like a hen's egg in outline) and drawn out to a long point, being what is called acuminate, and are often heart-shaped at the base. Their edges are serrated and the texture is smooth. The corymbs are from 2 to 6 inches across with from four to eight long-stalked, sterile, creamy white ray-flowers in early July, each from ½ to ¾ inch across and having four entire sepals. The seed capsule is short with ten prominent ribs. It is found over a wide range in the United States, extending from New York State to Iowa and south to Florida and Louisiana. It was introduced by Peter Collinson and apparently first flowered in his garden at Mill Hill in 1746.

The form usually found in gardens, *H. a. grandiflora*, has a large, pouffe-shaped head of sterile flowers. It was found wild in Ohio during the nineteenth century. Though frost-hardy, it is often rather a weakly grower unless given a rich, moist soil, kept well mulched. The flowers are of a rather greenish-white at first and apt to be too heavy for the stems, but the remarkably hardy nature of this species makes it useful for cold districts. When well grown, its foaming masses of white can be very attractive. (See Illus. 4.)

Other varieties are *H. a. oblonga*, with longer leaves rounded at the base, *H. a. oblonga sterilis*, a sterile-flowered variety of this type, and *H. a. australis*, a variety with leaves more prominently heart-shaped at the base and more coarsely toothed.

H. arborescens × *H. radiata* is the hybrid × **H. canescens**. It has little garden value, having few or no showy sterile flowers.

H. arborescens subsp. *discolor*. Small. This is a more bushy shrub up to 6 feet high with minutely downy young branches and ovate leaves with a very short point, bright green above and downy and greyish beneath. The corymbs are about the same size as those of the preceding species but with only very few small sterile ray-flowers. Figured Britton and Brown, *Ill. Fl. N.U.S.*, **2**: 231 (1897).

This species has a garden form with a ball-shaped head of sterile flowers, *H. c. sterilis*. It is rather similar to *H. arborescens grandiflora* but less effective. Indeed, *H. cinerea* is a sort of connecting link between *H. arborescens* and *H. radiata*. Its habitat is more southern, occurring no further north than North Carolina.

H. arborescens subsp. *radiata*. The distinguishing feature of this species is the more marked whiteness of the felted hairs on the undersurface of the ovate, tapered leaves and the fact that the 4- to 8-inch wide corymbs have more numerous, long stalked creamy-white ray-flowers, each often over an inch across in the better forms. It comes from the Carolinas. *H. radiata* has an upright habit, attaining about 3 feet usually, but reported up to 7 feet. Flowering in July it is one of the more attractive species of this subsection. It does not appear to be available in commerce in this country at the present time.

H. quercifolia Bartram, often called the Oak-leaved Hydrangea, has a still more southern range in the United States, occurring in Georgia, Florida, Alabama and Missouri. It reaches about 5 feet in height, making a dense, rounded bush. The young branches are covered with red fur and the leaves are lobed like those of the Oak. The flower-head is a panicle usually about 6 inches long with numerous attractive, round, white, sterile flowers, about an inch across, which ultimately turn purplish. The flowering time is usually from early July onwards. The autumn leaves colour well.

It is a pity that this handsome Hydrangea is just too tender to thrive in any but favoured gardens in the south and west of England. Even in these, it often just fails to be really decorative owing to the poor presentation of the flowers which usually droop among the leaves unless the season is an unusually favourable one which enables the more exposed branches to flower freely.

Subsection 2 ASPERÆ
Flowers white, blue or pink. Ovary completely inferior. Capsule hemispherical or turbinate, truncate at apex. Styles usually two. Seeds winged at the ends.

This is the subsection in which *H. Sargentiana* is probably the best known.

H. sikokiana Maximowicz is a little-known Hydrangea resembling *H. quercifolia*, but having fewer showy sterile flowers and more slender panicles. Its habitat is Koyasan, Japan. It does not appear to be in cultivation, but the fact that this

American type of Hydrangea also occurs in the Orient is a matter of interest. It is, of course, paralleled by the occurrence of members of the Cornidia section in both South America and in Formosa and other islands in the Orient.

H. involucrata Siebold is a Japanese shrub of 4 to 6 feet under good conditions, but more often seen only 2 feet high or less owing to all growth above ground being winter-killed. The young shoots, leaves and flower-stalks are covered with pale bristly down. The slender-pointed, ovate-oblong leaves are also margined with numerous fine bristle-like teeth. They are about 4 inches long and 2 inches wide and rough on the upper surface. The flower-head, about 4 or 5 inches across, is at first enclosed by about half a dozen large, broadly egg-shaped bracts about an inch long, covered with whitish down. The ray-flowers, projecting on long stalks, are ¾-inch to an inch across with from three to five sepals of a very pale blue or faintly pink-blushed white colouring. *H. involucrata* flowers late – usually in early August. The whorl of persisting bracts at the base of the corymb is a distinguishing feature and this species is usually much admired during the short time it is in bloom particularly when, grown in acid soil, the blue fertile flowers contrast prettily with the almost white ray-flowers. In the colder districts, however, the flowers often fail to open fully. It is not a very hardy species, suffering the loss of its top growth in hard winters. Though rather scarce, it is in cultivation in this country.

A variety with doubled and more numerous sterile flowers, slightly deeper in colour is *H. i.* 'Hortensis' Maxim. It is very free-flowering and effective, the pink flowers usually open well and cover the bush with a mantle of blossom. Though scarce, this handsome Hydrangea is available at specialist nurseries and it appears to be hardier than the 'type'. It was imported by T. Smith in 1906. (See Illus. 6.) There are said to be many attractive superior varieties of this species in Japanese gardens and their importation is highly desirable. 'Hortensis' runs at the root and its queer quadrupled flowers are puzzling, to say the least.

5 *Hydrangea aspera*

6 *Hydrangea involucrata* var. *hortensis* at Wisley

H. involucrata longifolia Hayata, another Formosan species, resembles the preceding but has long, lanceolate leaves, hairy seed capsules and entire, hairy sepals on the sterile flowers. Not in cultivation.

H. aspera Don is really a Himalayan species said to be distinguished from the preceding by the fringed, minutely toothed leaves which are paler beneath and rough to the touch, with long, soft, curly hairs. It is by no means easy to differentiate between these closely allied species, particularly as both also have Himalayan and Western Chinese forms which show local variation. Some of these are much superior to others in decorative value. Unfortunately the hardiness of this species is only sufficient for the southern and coastal counties and even then only in gardens where care is taken to install such fastidious flowering shrubs with proper respect, in well enriched soil. (Figured Hu and Chun, *Icones Pl. Sin.*, **1**: 28, t. 28 [1927]).

The form usually seen here is an upright shrub of 6–8 feet with about a dozen pale purplish ray-flowers in an inverted saucer-shaped head borne well above the foliage. A curious feature regarding the flower colour of the species of this subsection is that, with the exception of *H. involucrata*, this does not appear to be affected in the direction of greater blueness by aluminium. In acid soils, where *macrophylla* has vivid blue flowers, *H. aspera* usually remains purple. Blue-flowered forms of *H. villosa* are just as blue in limy soils where the garden hybrids have clear pink or red flowers. Owing to the very wide range of *H. aspera* in Nature it should be possible to obtain hardier forms than those available in this country at present.

G. Forrest recorded it as a shrub of from 6–12 feet with yellowish-white ray-flowers and pale purple fertile flowers growing amongst scrub in dry, open situations at elevations of from nine to ten thousand feet in north western Yunnan on the eastern flank of the Lichiang Range.

H. aspera oblongifolia Blume is described as having few or no ray-flowers but herbarium specimens show fine corymbs with a dozen or more large ones, each nearly 1½ inches across and these are of a cream colour tinged with pink, and the small fertile flowers are blue. A native of Java and Sumatra and sometimes growing as a climber, its appearance suggests that it is very near to *H. Kawakamii*.

H. aspera forma *Kawakamii* Hayata is, judging from the herbarium specimens sent back by Wilson, the largest and the most spectacularly fine of all the Hydrangeas of this subsection. It is a mighty, often climbing, species with huge, 18-inch, tawny-underfelted leaves and rather open corymbs over a foot across, with a dozen or more white or purple ray-flowers each measuring about 1½ inches across and having four serrated sepals. It was discovered in 1906 on Mount Arisan, Formosa, at 7,500 feet altitude, but Wilson evidently found still finer forms when he visited the area some twelve years later. It is described and illustrated in the *Journal of the College of Science, Tokyo*, **25:** t. 8 [1908]. It is unfortunately not in cultivation, *H. strigosa macrophylla* is the nearest species available.

H. aspera forma *fulvescens* Rehder is also related. The ray-flowers are described as having four white, orbicular-obovate, entire sepals. Not in cultivation.

H. aspera forma *glabripes* Rehder is related to *H. longipes* but has narrower leaves, still longer leaf-stalks and white flowers instead of purplish ones. To this species probably belongs a very fine form with about nine four-sepalled, boldly serrated, large, white ray-flowers which was sent back by Wilson from Kansu, Central China. It is very decorative but rather tender.

Hydrangea aspera forma *villosa* is, I suppose, what we should, in future, call what has heretofore been named *Hydrangea villosa*. It is the most renowned and, in its superior forms, one of the most spectacular of the pure species. But it is highly variable. Some forms are little better than the upright-growing, dull purple-flowered, spoon-leaved *aspera* of commerce. Even the best flowered form, for which I am indebted to the kindness of the late Sir Frederick Stern, V.M.H., of Highdown, has a somewhat untidy habit and the peeling, grey-brown puckered bark of the older stems is rather ugly. The flowers of the good forms are a pleasing violet-blue and the colour is unaffected by limy soil. It is not a plant that will grow just anywhere without care. Good soil conditions and extra water in hot summers are usually necessary. A height of 9 feet with a spread of six seems to be the usual size. The stems, leaf-stalks and flower-stalks are covered with dense, spreading curly hairs which are sometimes rust coloured. The leaves are fringed and minutely toothed and hairy on both sides. The undersurface is greyish-white. The plant was introduced from Szechwan, China, in 1908 by E. H. Wilson, and first flowered at Kew in 1915. This form was described by W. J. Bean as having toothed sepals. The best form that we have has them entire. (See Illus. 7.)

Indeed, the difference in garden value of the various forms is very wide and it is to be hoped that the finest kinds will be given distinguishing names and propagated as clonal varieties. *Villosa* is not, however, always very easy to strike from cuttings and so it may well be worth while to sow the seeds whenever these can be found, and to layer down any low branches. If the layers are watered but the parent is not watered, the layers root more quickly.

H. v. strigosior Diels is described by Rehder as a variety having 'the petioles and branchlets with few, or sometimes without, spreading hairs', and introduced into cultivation in 1905.

A form with a completely sterile-flowered, ball-shaped inflorescence of a pinkish colour was found by E. H. Wilson under No. 1473A, Packang, Central China. According to the label, plants were sent back to Messrs. Veitch. Apparently they were lost and I have, unfortunately, not been able to secure this beautiful variety.

7 *Hydrangea aspera* forma *villosa* at Highdown

8 *Hydrangea aspera* subsp. *Sargentiana* at Wisley

H. aspera subsp. **robusta** Hooker f. is a Himalayan species cultivated for nearly a century, but rarely seen. It is not very hardy, coming from Sikkim and Bhutan at elevations of from 5,000 to 8,000 feet. Bean (*Trees and Shrubs Hardy in the British Isles*) rightly described it as 'both in leaf and inflorescence one of the most striking of the Hydrangeas'.

It is a shrub of up to 15 feet in height, of spreading habit, with ovate-lanceolate leaves, conspicuously toothed and hairy. The corymb is nearly a foot across with twenty or more white ray-flowers up to 2 inches across with, usually, ovate, serrated sepals. The fertile flowers are blue. It was figured in the *Botanical Magazine* (as *H. cyanema*, t. 5038) in 1858 and Bean suggested that the plant shown was raised from seed sent home by Joseph Hooker a few years previously. The few plants now in cultivation are from a reintroduction.

H. aspera subsp. **robusta** forma *longipes* Franchet (*H. Hemsleyana Diels*) is a species of more low and spreading habit with lax branches and a similarly hairy character to the preceding. The entire inflorescence has a purplish cast at first and the small fertile flowers and the eight to nine sterile ray-flowers, which are about 2 inches across in the best forms, open a dull white, sometimes suffused with a purplish tint. The unusually long leaf-stalks are the characteristic feature. There are many worthless forms with wretched little flowers. But, judging from herbarium material, superbly flowered forms do exist. It was introduced from Central and Eastern China by E. H. Wilson for Messrs. Veitch of Exeter about 1901.

H. aspera subsp. **robusta** forma *Rosthornii* Diels is a western Chinese species introduced in 1908. It is a larger and more massive shrub, with similarly hairy growth and larger corymbs of flowers.

It was figured in '*Icones Plantarum Sinicarum*', Hu and Chun, **3:** t. 141.

Some forms have almost orbicular leaves 6 inches across and white ray-flowers with four boldly serrated sepals. The bush is sometimes 12 to 14 feet high. Though very scarce, this plant is still in cultivation in Britain. It is very near *H. robusta* if, indeed, it is distinct.

H. aspera forma *Rehderiana* C. K. Schneider is near *H. villosa* but differs in its curiously shaped leaves which are suddenly contracted near the pointed tips and have scalloped edges. It was collected in Western Hupeh, China, by E. H. Wilson (*Ill. Handb. Laubholzk*, **2:** 940 [1912]).

H. aspera forma *strigosa* Rehder is a Chinese shrub of 7 feet with erect growth. The branchlets are densely strigose, which botanical term means 'beset with appressed straight and stiff hairs' (A. Rehder). The leaves are lance-shaped and drawn out to a long point and rounded at the base. The undersurface is also strigose, especially on the veins, so too, are the leaf-stalks. The flower-head is a corymb about 5 inches across with whitish, or dingy pale purple ray-flowers, about an inch across. The sepals are broadly egg-shaped and entire or toothed. It flowers in August. Though interesting it is not a very decorative species except in its best forms.

A most beautiful Hydrangea is *H. strigosa* f. *sterilis* Rehder. It is apparently common in the wild, as many collectors have sent it back. The flower-head is globose and composed of mostly sterile flowers like that of the Hortensia. The buds are purplish but the flowers open pure white. Wilson collected this lovely variety at Fang Hsien, Western Hupeh, China, under No. 2390 and also, under No. 4902, from Mt. Omei. It is singularly unfortunate that this form is no longer in cultivation in this country, as even dried specimens present a picture of great charm.

The variety *macrophylla* (*H. aspera macrophylla*) is also a much finer form and has huge leaves and corymbs a foot across, with fourteen or more 2-inch ray-flowers with four serrated sepals. These are pale purple, turning red as they age, but remaining decorative for many weeks in autumn, making this one of the finest of the Hydrangea species. *H. strigosa* was discovered in Western Hupeh, China, by E. H. Wilson in 1901, but had been previously recorded by Augustine Henry.

It is almost equal in hardiness to *H. maritima*, but, like nearly all the Hydrangeas except that maritime species, requires light shade and shelter.

H. aspera subsp. **Sargentiana** Rehder is a large, coarse, rather leggy and gaunt shrub growing up to 9 feet high with densely hairy, rich velvety green leaves often 8 or 9 inches long, with purplish young shoots. The flat flower-head – about 5 or 6 inches across – has, usually, pale purple fertile flowers and whitish, slightly cupped, ray-flowers. The latter are not very large in proportion to the size of the plant, being about 1½ inches across, and usually having four or five entire, rather irregularly shaped, rounded sepals.

H. Sargentiana was first found in West Hupeh, China, by E. H. Wilson and introduced in 1908. Often described as rather tender, I think it is perhaps better described as fastidious as I have seen better plants in the wood at Wisley than in more open situations in Cornish gardens.

The gaunt, leggy habit may be avoided by early training and pegging-down, but this species definitely seems to require shade and a fairly rich, moist soil. A specimen or two in a shady place among Rhododendrons is a decorative feature, but it is of little value as a plant for ordinary garden use. It flowers in July and August. (See Illus. 8.)

Subsection 3 CALYPTRANTHE
Climbing by aerial rootlets, petals cohering at apex and falling off as a whole. Capsule truncate. Seeds winged all round, compressed.

In this section *H. petiolaris* is the best known.

H. anomala Don (*H. altissima* Wallich) is a related climbing species from the Himalayas introduced in 1839. Resembling *H. petiolaris*, it is slightly less hardy in the British climate. The four-sepalled ray-flowers are fewer in number, often only four, and the fertile flowers show a more yellowish tinge. Figured, as *H. altissima* Wallich, in *Tentamen Floræ Napalensis*,

t. 50 [1826]. It is in cultivation here and may be easily distinguished from the preceding species by its puckered leaves, more tender growth and sparser flowers.

George Forrest recorded it on the Tali Range in Western Yunnan as a shrub of 6–12 feet, with greenish-yellow ray-flowers, growing amongst scrub at 9,000 or 10,000 feet elevation.

Wilson sent back specimens of forms from Western Hupeh far superior to that in cultivation in Britain. These had from five to eight and more, rounded, white ray-flowers 1½ inches across and he described the corymbs as most agreeably fragrant.

H. glabra Hayata is merely a Formosan form of this species from 7,500 feet elevations on Mount Arisan. The four or more ray-flowers per corymb are ¾ inch across and have four entire sepals (*Jour. Coll. Sci. Tokyo*, **25, 19:** 89, t. 6 [1908]).

H. anomala subsp. **petiolaris** Siebold is a deciduous, climbing Hydrangea from Japan where it reaches to the tops of tall trees, clinging to the trunks by its aerial roots, like Ivy. The older stems have peeling, brown bark and the young shoots are green. The regularly and sharply toothed, thick, vivid green leaves are smooth on the upper surfaces and often tufted with down on the veins beneath. In the form of this plant available in Britain the somewhat irregular, spreading corymbs, opening in June, carry up to a dozen irregular white ray-flowers of various sizes protruding on long stalks from the margins. They are white and have, usually, four ovate sepals. The fertile flowers are dull white. The beautifully flowered example illustrated in Siebold's *Flora Japonica* (**1**: t. 54 [1840]) is unlike any form I have seen in this country and suggests that there may be more decorative forms in Japan than that available here, which is probably Siebold's *H. cordifolia*.

This climber has been with us since about the year 1878 but is seldom seen grown in a really decorative manner. As a wall plant it is surpassed by many with more vivid flowers, but when used to cover a tree-stump or hummock it forms a singularly picturesque low, bushy mound that is very attractive when in flower. It is also decorative when allowed to climb up a tree and, like Ivy (contrary to popular notions), does little, if any, harm to its host provided that it is not allowed to cover the top growth and foliage. In order to give such a climber a fair start it is advisable to plant it in a tub of prepared soil sunk in the ground suitably close to the tree-trunk. (See Illus. 9.)

When used as a wall plant, the effect is much improved by the addition of that beautiful climbing Nasturtium, *Tropæolum speciosum*, whose delightfully-shaped red flowers mingle with the later-opening white ones of the Hydrangea.

This species is less easy to strike from cuttings than others, but half-ripe shoots secured from a branch that has been cut back previously will usually give a fair proportion of rooted plants if placed under a hand-light in early July. Serpentine layering is an effective alternative method of propagation.

Monsieur Henri Cayeux, the eminent French breeder of Le Havre, informs me that he successfully crossed × *H. macrophylla* var. *rosea* Veitch with *H. petiolaris* and produced a curious plant which he named × *H. hortentiolaris* and exhibited in Paris about 1922. This hybrid was plainly intermediate in character, but unfortunately all the plants were destroyed by the intensive bombardments to which the unhappy city of Le Havre was subjected during the late war.

H. p. cordifolia. A variety with smaller, more numerous and regular ray-flowers and leaves heart-shaped at the base, described and figured, but only from one small piece of material, by von Siebold (*Flora Japonica*, **1**: 113 t. 59, fig. 2 [1840]). The leaves of the common form available in this country are slightly variable in shape and the flowers are more like those depicted for var. *cordifolia* than those depicted for the 'type' in the same work (t. 54).

Subsection 4 PETALANTHE

Inflorescence corymbose. Petals of fertile flowers pink or blue and mostly persistent. Ovary half or partly superior. Capsule ovoid, narrowed at apex into, mostly, three styles. Seeds not winged or with very short wings.

9 *Hydrangea anomala* subsp. *petiolaris*

10 *Hydrangea scandens variegata* at Grayswood Hill

This is the subsection in which *H. maritima* (in its globose-headed form 'Joseph Banks') is probably the best known of the pure species.

H. hirta Siebold, a Japanese species from the mountain valleys of Japan, is a shrub of 3 to 4 feet with nettle-like leaves and corymbs composed only of small fertile flowers. It appears to have little or no garden value.

H. scandens Linnæus (*H. virens* Siebold and Zuccarini). This Japanese species has fine forms bearing nice little corymbs, with about three, inch-wide, white ray-flowers, at every node all along the arching branches, but I do not think that these good forms are in cultivation in Britain. This is most regrettable, as this species would probably prove of great value in hybridisation, owing to its amazing freedom of flower. It is, indeed, the most floriferous Hydrangea species I have seen.

The only form we have here is an inferior Japanese garden clone, grown only as a curiosity, with a weakly habit and small lance-shaped leaves variegated with dull purple and yellow. It does not flower very freely and the little corymbs have only two or three, three-sepalled, white ray-flowers of no great decorative value. (See Illus. 10.)

H. scandens subsp. **liukiuensis** Nakai, a species from the Liu Kiu Islands, is near *H. scandens* but has thicker leaves, longer anthers and no sterile ray-flowers (*Bot. Mag. Tok.*, **25**: 63 [1911]).

H. scandens subsp. **chinensis** Maximowicz (*H. yayeyamensis* Koidzumi) is a more hard-wooded species with smooth leaves and branches. Described by Rehder as having few or no ray-flowers, herbarium material indicates many forms with fine large white ones and fertile flowers having blue anthers, and, having seen these, I am inclined to the opinion that the ray-flowers may have dropped off some of the earlier dried specimens. The long stalks of the ray-flowers are exceedingly slender and fragile. It is apparently a variable species, occurring over a wide area in China and the Liu Kiu Islands. So far as I know, it is not now in cultivation in Britain but the best forms appear to be highly desirable. In the botanical revision of this genus it has been decided that this species covers *H. umbellata* Rehder and also some forms now described as coming under *H. yunnanensis*. Other forms of the latter, with fine large flowers may be considered as merely geographical forms of *H. Davidii*. The Formosan species formerly often described as *H. chinensis* probably comes under *H. macrosepala* Hayata.

H. chinensis is distinguished by its hard-wooded twiggy character and dark brown, close bark, the stems being more permanent and less successional than those of other allied species. Having such a wide range, it should be possible to find very hardy forms and these would be a useful addition to the ranks of flowering shrubs and of great value to the hybridist.

I believe that the cultivar 'Lanarth White' (p. 125) is a hybrid of the subspecies *chinensis*.

H. scandens subsp. **chinensis** forma *Lobbii* Maximowicz is a superb species from the Philippines. Although placed in this subsection, it has, to my eye, the superficial appearance of being akin to the cultivated forms of the Petalanthæ. It is described as occurring 'in mossy forests on most of the higher mountains at from 1,000 to 2,400 metres elevations from Luzon to Mindanao'. The four or five fine white ray-flowers are each 2 inches across and have four serrated sepals. Unfortunately it is not in cultivation.

H. scandens subsp. **chinensis** forma *Davidii* Franchet is a tall shrub up to 6 feet in height with the young branches downy at first, and oblong lance-shaped leaves, wedge-shaped at the base, with serrated edges. They are of rather a yellowish-green and slightly furry on the veins underneath. The flower-head is diffusely branched and loosely formed, roughly bun-shaped and about 6 inches or more across. The fertile central flowers are blue or pink and the sterile ray-flowers are about 1½ to 2 inches across, and have four rounded oval, entire sepals white in colour. The upper sepal is much smaller than the others and the seed vessels are cup-shaped. This species comes from Western China and was introduced in 1908, having been originally discovered in Moupin by the Abbé David in 1869. Unfortunately this interesting and attractive Hydrangea does not appear to be in cultivation in Britain at the present time.

H. scandens subsp. **chinensis** forma *Pottingeri* Prain from 4,100 feet elevation in the Kachin Mountains in Burma is interesting as it has the appearance of being a Burmese form of *H. japonica*, a species which has also Chinese and Himalayan counterparts. The ray-flowers have three serrated sepals rather smaller and less extreme in shape than those of the Chinese and Japanese forms. Not in cultivation.

H. scandens subsp. **chinensis** forma *umbellata* Rehder, a handsome species with slender, dark brown stems, markedly serrated lanceolate leaves and numerous corymbs about 5 inches across, each having three to five bluish-white ray-flowers each an inch or more across, comes from S. Anhwei and Kiangsi, China, and is a shrub 4 or 5 feet high growing at elevations of 2,800 feet. Described as near *H. scandens*, it seems to me to be nearer a good form of *H. chinensis* (see Hu and Chun, *Ic. Pl. Sin.*, **3**: t. 131 [1933]).

H. scandens subsp. **chinensis** forma *yunnanensis* Rehder is a very handsome species from Yunnan, intermediate between *H. chinensis* and *H. Davidii* in character with numerous large, ray-flowers often 2 inches across and with leaves very similar to those of × *H. serrata*. This looks like a valuable plant for the hybridist.

It was apparently this species that G. Forrest recorded (as *H. chinensis*) as a semi-scandent shrub of 4 to 8 feet with white ray-flowers and cream-coloured fertile flowers with purple anthers, growing in dry, open situations among scrub on the eastern flank of the Tali Range in Western Yunnan.

H. scandens subsp. **chinensis** forma *glabrifolia* Hayata is a Formosan species described as near to *H. chinensis*, but it is obviously inferior, having few or none of the showy ray-flowers. The habit is upright and the leaves are pointed and minutely toothed. (Hayata, *Icones Plantarum Formosanarum*, **3**: 106 [1900]).

H. scandens subsp. **chinensis** forma *macrosepala* Hayata, another Formosan species resembles *H. chinensis* but has purplish brown stems, smooth, almost entire, merely notched, lanceolate leaves and very much larger flowers. There are from four to six white ray-flowers per corymb and these are over 2 inches across and have four unequal sepals, in shape obovate-orbicular and slightly serrated at the edges. (*Ic. Pl. Formos.*, **3**: 108 [1913]).

H. scandens subsp. **chinensis** forma *obovatifolia* Hayata has obovate leaves and a corymb about 1½ inches by 2½ inches with curiously shaped sterile flowers having four broad sepals only ⅝ inch long. Otherwise it resembles *H. glabrifolia* (*Icones Plantarum Formosanarum*, **3**: 108 [1913]).

H. scandens subsp. **chinensis** forma *angustisepala* Hayata is, apparently, a Formosan counterpart, described as being more downy and having more elongated sepals.

H. scandens subsp. **chinensis** forma *pubiramea* Merrill, another Philippine species, is described as a shrub 6 feet high with the young branches, leaf-stalks, flower-stalks and under-surface of the leaves covered with appressed brownish hairs. The branches are greyish-brown and the leaves, of the texture of stiff paper, are oblong or oblong-obovate (like a hen's egg in outline with the stalk at the small end). In colour they are 'olive grey when dry' and slightly glossy. In size they are about 4 inches long by 1½ inches wide. As is common with botanical descriptions based on dried fragments, few data are given regarding the most horticulturally interesting portion of the plant – the flowers. We are, however, told that the inflorescence is terminal, peduncled and pubescent and also that the flowers are only borne at the top of the plant. *H. pubiramea* is said to be akin to *H. Lobbii* but is distinguished by being hairier. A variety, *parvifolia*, is smaller in all its parts. The habitat is given as Luzon, 'in wet forests at an elevation of 300 metres or over'. It is not in cultivation.

H. scandens subsp. **chinensis** forma *formosana* Koidzumi is quite distinct, having long pointed, narrow, willow-like leaves very closely set on the stems and large, handsome, white ray-flowers.

H. scandens subsp. **chinensis** forma *stenophylla* Merrill and Chun has no ray-flowers and very long and narrow, willowy leaves. The fertile flowers are described as green (*Sunyatsenia*, **1**: 58 [1930]). There are quite a number of these distinct, queer, willowy-looking species and it may be found necessary to instigate a separate species for them. Not in cultivation.

Subsection 5 HETEROMALLÆ

Inflorescence corymbiform and convex, or paniculiform. Fertile flowers white. Petals of fertile flowers quickly deciduous. Ovary semi-superior. Capsule ovoid, narrowed at apex into, mostly, three styles. Seeds winged at the ends.

This is the subsection in which *H. paniculata* is probably best known.

H. paniculata Siebold. This very hardy Hydrangea is probably the most commonly grown as an outdoor shrub in the colder districts of Britain. In favourable conditions it will grow up to 15 feet high and in Japan has been reported as a tree of nearly 30 feet. The wild plants vary very considerably, some having thin little panicles hardly worth looking at, others have many broad-sepalled, large ray-flowers of a rich cream colour often with a little rosy-coloured pip or budlet in the centre. Such superior forms are really more decorative in the garden landscape than the variety usually grown – var. *grandiflora*. By pruning and thinning the latter, enormous lumps of flower can be produced. The panicle, almost entirely composed of the large sterile flowers, sometimes attains 18 inches high and 12 inches through at the base. (See Illus. 11.)

H. paniculata is a native of Japan, the island of Saghalien and China. The framework of the shrub is composed of stout strong branches and the leaves are elliptic and drawn out to a sharp point. The sterile flowers, set around the outside of the panicle in tiers, are creamy white with four entire, elliptic sepals. Later they change to a purplish-pink colour. The 'type' flowers in August but a variety found in America – *H. p.* var. *præcox* – flowers in early July, with longer and narrower sepals to the ray-flowers and a shorter, smaller panicle. It is a most useful midsummer shrub. Award of Merit 1956. It comes true from seed and has a strong scent of mixed horse and Chanel No. 5. (See Illus. 12.)

H. p. floribunda is a clone whose every shoot and sideshoot terminates in an inflorescence of mixed fertile and sterile flowers of a creamy-white with a small red eye. It is superior to *grandiflora* as a flowering shrub but not, perhaps, for cut blooms. (See Plate A, facing p. 82.) Award of Merit 1953.

Monsieur L. Foucard of Orléans succeeded in 1910 in breeding a most remarkable hybrid between × *H. macrophylla* var. *rosea* Veitch and *H. paniculata*. This hybrid had pink flowers and clearly showed its parentage in its intermediate character. The plant was exhibited before the Société Nationale d'Horticulture in 1912, when it received a Certificate of Merit. An illustration appeared in *Revue Horticole*, p. 106, 1912. Most unfortunately this unique plant was lost during the ensuing War and with it, apparently, the opportunity to found a race of ultra-hardy Hydrangeas with coloured flowers. This species is best propagated from hardwood cuttings in winter.

A superior seedling 'Mt. Everest' has a huge flower-head of blushed white flowers, some fertile ones, on a much stronger bush than the weakly 'grandiflora'.

H. paniculata forma *Schindleri* Engler from Kiangsi, E. China, is a very fine species, judging from herbarium speci-

11 *Hydrangea paniculata grandiflora*

12 *Hydrangea paniculata* var. *praecox*

13 *Hydrangea heteromalla* at Abbotswood

mens. It is remarkably free-flowering, seven corymbs being borne on a foot-long length of stem. Each corymb carries four or five white ray-flowers, each 1½ inches across and with long, pointed sepals. Rehder put this species down as a form of *H. paniculata*. It is very close to *H. umbellata* Rehder.

H. heteromalla Don (*H. vestita* Wallich) is a shrub up to 9 feet in height with close, olive-brown bark. The leaves are narrowly oval or lance-shaped, drawn out to a long point and wedge-shaped at the base. They are usually about 6 inches in length and densely bristled and toothed at the edges. The upper surface of the leaf is smooth but the undersides are covered with white down. The leaf-stalks are red. The somewhat thin corymbs of flowers are about 6 inches across and the flower-stalks are downy. The sepals of the white, inch-wide ray-flowers are elliptic or oval. *H. heteromalla* flowers, like most Hydrangeas, in July or August. This Hydrangea is in cultivation in England and is quite a decorative shrub with its white underfelted leaves, red stalks and white flowers. It is, however, less hardy than *H. Bretschneideri*. Introduced from the Himalayas in 1821★ it has since been found in Central and Western China and forms from these localities were introduced by E. H. Wilson more recently. (See Illus. 13.)

H. heteromalla varies widely; some forms are not worth growing, others are very beautiful, with as many as thirty inch-wide, shapely, round-sepalled, straw-yellow ray-flowers to a corymb. A form sent back from open situations in the Lichiang Range, N. W. Yunnan, China, at an altitude of 10,000 feet is described as having the sterile flowers of a greenish-yellow with the undersurfaces purplish-blue. *H. Khasiana* Hooker fils, sent back by Forrest from 10,000 feet elevations on the Mekong-Salwin Divide, whence come the Rhododendrons in their greatest numbers, is said to be *H. heteromalla* var. *mollis* Rehder. No shrub in the Demonstration Gardens seeds itself so successfully as this species.

★ Figured, as *H. vestita* Wallich, *Tentamen Floræ Napalensis*, t. 49 [1826].

H. heteromalla forma *Bretschneideri* Dippel is a sturdy, hardy, bushy species up to 10 feet high. The bark of the matured branches peels off in flakes and the leaves are ovate, and pointed, smooth and dull on the upper side, with down on the veins beneath. The flat corymbs are about 4 to 5 inches across with from about six to nine dull white ray-flowers each about an inch across. The flower-stalks of the corymb are covered with bristly down.

H. Bretschneideri comes from China, having been introduced from mountain districts near Peking about 1882 by Dr. Bretschneider. It makes quite an attractive specimen shrub or small tree depending upon whether the position is open or shady. It is often in flower by late June but it does not last in beauty so long as most Hydrangeas. The ray-flowers soon turn to unattractive purplish and greenish tints and thus it is not a shrub for a prominent position. In any event it requires a deep rich and moist soil to do well, though it is a hardy species that should be more generally planted. It succeeds *Viburnum tomentosum Mariesii* with a similar flower effect and is available at specialist nurseries.

H.B. var. *glabrata* is similar but smooth, not hairy.

H. heteromalla forma *mandarinorum* Diels is a distinct, smaller species that is quite attractive in habit, with many small, white ray-flowers, each ½ inch across with four even, pointed sepals. It has downy young branches and minutely serrated, velvety, elliptic-acuminate leaves with the veins conspicuously paler in colour. It was obtained from Nan ch'uan, Central China, and also occurs, apparently, in Yunnan. It is in cultivation in this country. (See A. Engler, *Botanische Jahrbucher*, **29:** 372 [1901]).

H. heteromalla forma *xanthoneura* Diels. This tall tree-like species, with chestnut-brown bark, reaches 35 feet in this country.★ The large leaves are bright green, elliptic-oblong and abruptly pointed. The corymbs of white flowers are about

★ A specimen at Monreith, Wigtown, has exceeded this height.

6 to ·9 inches across with numerous ray-flowers up to 2 inches across with rounded oval sepals.

H. xanthoneura comes from China whence it was introduced for Messrs. Veitch by E. H. Wilson about 1904. It is quite a decorative species, but care has to be taken to give the shrub a fairly open position. If crowded up with other plants it grows very 'leggy' and the flowers are too high up to be seen. It flowers in late June or early July.

H. xanthoneura Wilsoni is distinguished by paler and less peeling bark and is not so tall.

H. heteromalla forma *hypoglauca* Rehder, a native of Western Hupeh, China, is described (*Pl. Wils.*, **1**: 26 [1911]) as a closely related species with the leaves bluish-white on their undersides. It was introduced by E. H. Wilson about 1901 and grows up to 9 feet high. W. J. Bean stated (*Trees and Shrubs Hardy in the British Isles*) that the leaves are ovate to ovate-oblong with a short, slender point and rounded base, finely toothed and about 3 to 5 inches long and 2 inches or more wide, dull green above and pale and rather glaucous beneath. The flower-head is a roughly convex disc, appearing on the ends of the current year's shoots in June. The ray-flowers, up to twenty and an inch across, have four unequal, rounded, white sepals. The small fertile flowers are white. This species was first discovered by Prof. Augustine Henry in a locality near Ichang, Central China.

H. heteromalla forma *pubinervis* Rehder appears to be near *H. xanthoneura* but inferior in flower, having about six to twelve small, scattered, white ray-flowers, but it has notably hard, black, woody old branches. It is not in cultivation.

H. heteromalla forma *dumicola* W. W. Smith. This is another large species from Western China with branches and flowers rough to the touch, and large leaves of a pale green, with appressed white hairs on the underside. The rather similar flower-head is about 10 inches across.

H. heteromalla forma *macrocarpa* Hand.-Mzt. has ray-flowers about an inch across with four very unequal broadly ovate sepals. Otherwise it resembles *H. Bretschneideri*. Not in cultivation here.

Subsection 6 MACROPHYLLÆ

× **H. macrophylla** (Thunberg) Séringe. This name, being based upon Thunberg's specimen which represents the Japanese hybrid clone known as var. 'Otaksa', covers the whole race of garden hybrids of both the Lacecap and Hortensia sections having the 'blood' of the wild maritime species as well as that of the woodland species in their composition.* I have carefully examined Thunberg's original specimen and there seems no room for doubt. (See Illus. 1.) Wilson sent back specimens of 'Otaksa' from the temple gardens on Mt. Omi.

The leaves of 'Otaksa' are almost orbicular and one may wonder from which of the ancestral species this trait is derived. 'Otaksa' also appears to be less hardy than any of the known ancestral species, though this is largely due to the untimely nature of its growth. An unnamed species from 'Tungtze to Tsunyi', Kweichow, China, sent back by Handel-Mazzetti, looks suspiciously like having had a part in the ancestry of 'Otaksa', possibly *H. stylosa*.

The blood of this southern, inland species would account for the greater tenderness of some Hortensia varieties than the known ancestral species. Indeed, these varieties usually have a strong look of 'Otaksa' about them. Examples are 'Munster', 'Carmen' and 'Europa'. The characteristics of the other species involved, as they affect the Hortensias, are dealt with at length on pages 142 and 143.

---

* I may say that I have, since 1950, grown a large batch of seedlings of the second generation from the variety 'Mariesii' and my hypothesis of the hybrid origin of *Hydrangea macrophylla* is amply proved by examples of all the suggested parental wild species duly appearing among the seedlings. This collection will be kept available for study by students.

H. Maritima Haw.-Booth. I propose this name for the pure wild maritime species discovered in 1917 by E. H. Wilson on the shores of Japan and named by him (*Journal of the Arnold Arboretum*, **4**: 233–46 [1923]) **H. macrophylla** var. **normalis**. As the name *H. macrophylla* belongs to Thunberg's hybrid race it cannot be used for a wild species which is only one of the ancestors of that hybrid race.

E. H. Wilson described this species as follows:

This is the wild type and is distinguished by its flat corymb of hermaphrodite fls. with a few outer sterile pink ray-flowers each from 3 to 5 cm. in diam. The fruit is stout, yellow-brown and erect, narrow-ovoid, 6–8 mm. long, 3–4 mm. wide, ribbed, crowned by 3 diverging woody styles from 1–3 mm. long; pedicels rigid 7–10 mm. long. The habit, the character of the shoots and the shape and texture of the leaves are all similar to those of the sterile form.★

The number of ray-flowers with petaloid sepals is variable and so, too, is their size, and their colour is of varying shades of pink to rosy red, occasionally bluish or white.

This is a littoral plant abundant on the coasts of Oshima or De Vries Island and the Boshu peninsula on the east coast of central Hondo and not far from the port of Yokohama. It is also plentiful on Hachijo Island, a volcanic island south of Oshima and on Aoga-shima, another island just south of Hachijo.

The wild plant is a shrub from 1–3 m. tall with many stems forming a broad bush which occurs either singly or many together forming a dense thicket.

Its discoverer described the plant as growing close to the very shore, though sometimes among lava rocks above, but always exposed to the full influence of the sea.

Seeds and herbarium specimens were sent to the Arnold Arboretum, Jamaica Plain 30, Boston, Mass., U.S.A., but, I learn from a letter from the Assistant Curator, there are no plants of this species at the Arboretum now.

The Japanese name for this species is 'Gaku' or 'Gakubana'.

★ As the 'sterile form' is not specified this does not convey very much. The character of the shoots and the shape and texture of the leaves vary immensely in the globose-headed hybrid varieties. However, I presume Wilson had 'Joseph Banks' in mind. As, however, he considered the continental-bred Hortensias conspecific, there is no certainty on this point.

Apart from a small importation by the writer, *H. maritima* has not been imported into Europe on purpose, so far as I can discover. Its presence here is due to a plant of its globose-headed sport, 'Joseph Banks', having reverted, as a branch-sport, back to the wild species. For many years I had grown the latter, as a clone of puzzling origin, under the 'kennel name' of 'Seafoam'. I had recognised and described its evident close affinity to the maritime species. It was only later however, that I found this plant, *H. maritima*, actually occurring as a branch-sport, or reversion, on plants of 'Joseph Banks' in the Shanklin district of the Isle of Wight. The mystery was thus at once solved beyond doubt.

As a garden plant our clonal form of *H. maritima* is only of value near the sea, and there its variety 'Joseph Banks' is generally preferable. The wild species seldom flowers freely inland, indeed in ten years I have only secured three or four corymbs. In coastal gardens it flowers quite freely but requires moisture at the root and even light shade to look its best. This form of *H. maritima*, propagated as a clone in this country distinguished as *H. m.* var. 'Seafoam', is a shrub of even more markedly tip-flowering habit and less vigour than 'Joseph Banks'. In the young state the leaves are narrower, more lance-shaped, less thick and shiny and of a greyer green. In size they may be up to 7 inches long and 4 inches wide. In shade, and with maturity, the leaves become broader, more rounded and shiny and of a more vivid green. Thus it may be assumed that in all probability 'Joseph Banks' originally occurred as a branch-sport on a lusty and vigorous mature specimen of *H. maritima*.

As E. H. Wilson described the wild plants as variable and the form grown in this country is of the one clonal variety, it is proper that this should have varietal rank and I therefore propose for it the name *H. maritima* var. 'Seafoam'.

'Seafoam', whose foliage has been described above, has a somewhat irregular, flat inflorescence usually about 6 inches across. The ray-flowers, normally from about six to eight in number, are variable in size and symmetry. Some are only partly formed. A typical fully formed ray-flower is about an

14 *Hydrangea maritima* var. 'Joseph Banks'

15 *Hydrangea maritima* Haw-Booth

inch across, has four sepals of almost pure white in a shady position, but flushed with pink or blue, depending on the acidity of the soil, in a more open position. If the plant is starved, the edges of the sepals are slightly serrated. If in a moist rich soil the sepals swell out to an almost entire and more rounded form. In brief, a ray-flower of 'Seafoam' closely resembles a floret of 'Joseph Banks'. The numerous fertile flowers in the centre of the inflorescence are pink or blue. Near the sea 'Seafoam' flowers on the end of each main shoot, but inland the side growths do not often flower, unless well ripened. Thus it is even more shy-flowering than 'Joseph Banks'.

H. maritima is not so fastidious, as to soil, as most Hydrangeas of this series and vigorous plants may be seen growing on dry cliff-tops in chalk-rubble and heavy clay. On the other hand, although I have seen inviting earth cliffs below massed plantings of 'Seafoam' which must have sent showers of seed down every year, I have never found self-sown seedlings of this species. Possibly the alkalinity of the soil was the cause, but I have observed that the numerous self-sown Hydrangea seedlings, which appear in my own and other gardens, are all of the woodland species type, although fertile-flowered hybrid varieties of the maritime type have equal opportunities for seeding themselves.

Probably this species does not reproduce very readily from seed and this would account for the Japanese having used the woodland species dominantly in the production of their garden varieties. Though all the woodland species have been commonly grown in Japanese gardens since very ancient times, only the hybrids 'Azisai', 'Mariesii', 'Otaksa' and 'Rosea', containing some of the blood of the maritime species, appear to be much grown in Japan, and this only comparatively recently.

A plant of this wild maritime species obtained from Japan resembles 'Seafoam' but has slightly broader and shinier leaves. The stalklets of the flower-head are more cramped and upright and the six or seven white ray-flowers are poorly shaped and on slightly longer individual flower-stalks. The stems are unspotted, being of plain green. The central fertile flowers are bluish. The bush is 5 feet high and of little ornamental value.

On the other hand, the wild woodland species *acuminata*, *japonica* and *Thunbergii* are all decorative garden-worthy shrubs and thus we see clearly why the Japanese took these into their gardens but never thought of taking in *H. maritima*. The clone 'Seafoam' is shown in Illus. 14. It has produced a sport known as 'Maculata' which carries splashes of white on some leaves but is rather apt to revert to plain green.

Two other variegated varieties may be mentioned here. 'Tricolor' is a sport of the cv. 'Mariesii' with three colours on the leaves – white, pale green and deeper green. 'Quadricolor' is the most striking of the variegated sorts. The leaves are cream, pale green, deep green and vivid yellow at the edges. It has a pale pink lacecap type flower-head and seems reasonably hardy, though extra palatable to slugs.

× *H. m.* var. 'Joseph Banks' (*H. hortensis* Smith); lacking a varietal name that is not confused as to its exact application, the foregoing is proposed. This clonal variety imported from China by Sir Joseph Banks in 1789, and still common in British seaside gardens, appears to be merely a globose-headed sport of the pure species. It has very stout massive shoots, very large, thick, shiny, wedge-pointed leaves and flowers only from the terminal buds. When these are winter-killed the plant does not flower. The corymb is unusually large and bun-shaped rather than spherical in shape, and the sterile flowers of which it is composed have pointed, usually entire but some-times slightly serrated sepals of a pale pink colour. With aluminium feeding, or after some years' growth in acid soil, the flowers are pale hyacinth blue. Characteristically, a faint darker blue streak, as though made with a pen, may be found on some of the sepals.

'Joseph Banks' was used by French breeders in the produc-tion of the Hortensias and its characteristics may be seen in such varieties as 'Mme de Vries', 'Oriental', 'Mme Legou', 'Monsieur Ghys', etc. Further information on this point is given in the chapter on breeding.

No Hydrangea makes a larger or more massively impressive hard-wooded shrub in the favoured coastal districts of Britain and strong hedges may be seen braving the winds of the Land's End district where such trees as Oaks and Sycamores are dwarfed to stunted bushes by the gales.

Owing to its weak flower colour and shy-flowering nature inland, it is generally less effective than the hybrids coming under *H. macrophylla*, but the massive permanent wood is a valuable asset in a flowering shrub. (See Illus. 14.)

H. macrophylla subsp. **stylosa** Hook. F. and Thoms. has the appearance of being a Himalayan form of *H. japonica*. Resembling it closely, the ray-flowers are 2 inches across with bold serrations at their edges. The fertile flowers are blue or pink, the sterile flowers white. Not in cultivation.

H. macrophylla subsp. **stylosa** forma *kwangsiensis* Hu is a thin, weedy species with thin, upright corymbs with only two or three small white ray-flowers. The fertile flowers are bluish. It belongs to the same group as the Japanese woodland species. Not in cultivation.

H. macrophylla subsp. **stylosa** forma *indochinensis* Merrill, a species from Indo-China, has about five sterile flowers per corymb. These are whitish in colour, borne on long stalks and about 1½ inches across. There are four ovate, rounded sepals. The leaves are thin and without down (Hu and Chun, *Ic. Pl. Sin.*, **3**: 35 t. 135 [1933]). It is described as near *H. kwangsiensis* Hu, but appears, from dried specimens, to be much superior in decorative value. Not in cultivation.

H. macrophylla subsp. **Chungii** Rehder is a rare and valuable Hydrangea described as having purple flowers, which, if it really belongs here, makes it almost unique in this subsection, which is of special interest from the horticultural point of view, as its members may be expected to breed with the cultivated races × *H. macrophylla* and × *H. serrata*. Rehder described it as related to *H. Mœllendorffii*, but I must say that, having examined both, I can see no resemblance. It is a hairy species with attractive compact corymbs about 5 inches across. The flower-stalks are very downy and there are about six fine, solid ray-flowers, each over an inch across and having four, even, orbicular sepals. The leaves, which are oblong-acuminate, have serrated edges and there is much down on the veins underneath. This species was found near Fort Yenping, Province of Fukien, China, at an altitude of 240 metres, in 1924. (See *Journal of the Arnold Arboretum*, **12**: 69 [1931].)

H. macrophylla subsp. **serrata** (Thunberg) Séringe. The original type specimens shown on pages 20 and 21 seem to me, now that my knowledge has been fortified by having obtained the wild plants alive from Japan, to show *H. acuminata* on Illus. 2 (left) and *H. japonica* (Illus. 2, right). Indeed even these dried old specimens are so obviously different from one another in both leaf and flower that in my original edition, thirty years ago, I could only think of them as hybrids. If, as I believe, these Hydrangeas are still speciating, in increased adaptation to their environments, a time must come some day when their species status will have become indisputable. The garden forms of the above are described on pp. 139 and 140. Now, Thunberg must have used the adjective *serratum* – meaning serrated – to describe as a distinguishing feature, the *flowers*, not the leaves because all the Hydrangeas had serrated *leaves* anyway. One may well wonder if the *acuminata*-like specimen with its unserrated flowers is really in its right place.

H. Thunbergii Siebold (**H. macrophylla** subsp. **serrata** McClintock). This is a dwarf species seldom more than 2 feet high. The slender young shoots are neither spotted nor streaked but suffused with red as are also the leaf-stalks and leaves which are inclined to be obovate and acuminulate, that is to say, outlined like a hen's egg with the bigger end outward and then ending abruptly in a short point. The corymbs open early and are about 3 or 4 inches across with from seven to twelve small, rounded ray-flowers each about ½ inch across. They have usually three almost orbicular, entire sepals of an unusually rich pink (H.C.C. Rose Madder 23–1) or purplish blue in acid soil. It has more than twice the amount of colour carried by any of the other species involved in the production of the garden Hortensias. *H. Thunbergii* is common in the nursery trade and was first flowered by Messrs. Cripps of Tunbridge Wells in 1870.*

This species appears to be found in a wild state among the wooded hills of Japan along with the others. From the *H. Thunbergii* strain are derived the hybrid Hortensias with small, round, deep coloured flowers. It will be noted that the deep coloured ones are all dwarfs, having comparatively small individual flowers. (See Illus. 17.)

Now *H. Maritima* has a very limited distribution in the wild, consequently there is not much variation within the species. But what was known as *H. serrata* which, in my view, really included *H. japonica*, *H. acuminata* and *H. Thunbergii*, inhabits the Japanese islands from one end to the other – from farthest north to farthest south. There is thus ample scope for the tremendous variation which I believe offers grounds for specific distinction. Horticulturally we shy at lumping into one species a slender little shade plant with red stems and narrow-sepalled white flowers that turn vivid red only when exposed to sunlight, a stalwart blue-flowered bush whose flowers change to pink only in limy soil and a pink-flowered dwarf bush.

H. acuminata Sieboldt (H. macrophylla subsp. serrata McClintock).

This species is characterised by a roughly disc-shaped, slightly domed inflorescence having about six ray-flowers each with four ovate, entire, even-sized sepals of a pale pink or, very readily with a relatively small degree of soil acidity, of a pale blue colour. The leaves are prominently serrated and are thicker and more ovate than those of the preceding. They often turn up slightly at the edges, giving a concave shape. The young shoots are dusky green with a few dark brown streaks. In size the shrub is slightly smaller than *H.*

---

* *Gardeners' Chronicle*, p. 1698 [1870].

† Siebold and Zuccarini, *Flora Japonica*, 1: 110, t. 56, 57 (1840).

16 × *Hydrangea japonica* var. 'Grayswood'

17 *Hydrangea Thunbergii*

japonica. In Japan it is said to have a more northerly distribution and in cultivation it proves more cold-resistant, and a little stronger in constitution. From this species are derived the free-flowering, slender-branched, readily bluing hybrid Hortensias of the 'Rosea' (Veitch not E. H. Wilson) and 'Vibraye' types. *Acuminata* flowers very early, in early June.

This is, I think, an even more garden-worthy wild species than *H. maritima*, being smaller, hardier and brighter in flower colour. Varieties derived from this species are described on p. 140 and illustrated in Illus. 18.

H. japonica Siebold (included under **H. macrophylla** subsp. **macrophylla** McClintock). This woodland species first described as a pink-flowered garden variety probably containing a trace of hybridity, by von Siebold,★ is best known in cultivation in the beautiful lace-flowered variety *H. japonica macrosepala* Regel.† (See Illus. 19.) The characteristics of this species are as follows. It makes a tall, very slender-twigged shrub up to 6 feet in height and breadth, in shade and moist soil. The corymb is often peculiar in shape, being formed like a U with the centre of the interior part, devoted to fertile flowers, much lower than the outer part on two sides. The fertile flowers are white, opening from very pale pink buds. The sterile ray-flowers are usually only four in number but very large compared with those of allied species, being 2 inches across. They are pure white and have, normally, three boldly serrated sepals. The lanceolate leaves are brilliant green, evenly tapered to a point and have wide but shallow-cut teeth, sometimes reduced to a mere minute notch. The young stems are green with a few faint, pale reddish-brown spots. From crosses with this Hydrangea are derived the bulk of the white-flowered Hortensias. If much exposed to sunlight the ray-flowers turn crimson, but never blue, however much free aluminium may be present in the soil. Crossed with *H. Thunbergii* this Hydrangea produced the × *H. serrata* variety × *H.s.* 'Intermedia'.

★ Siebold and Zuccarini, *Flora Japonica*, **1**: 107 (1840).
† Regel, *Gartenflora*, **15**: 290, t. 520 (1860).

18 *Hydrangea acuminata* 'Bluewave', awarded a first-class certificate by the Royal Horticultural Society

19 *Hydrangea japonica* var. *macrosepala*

Since the third edition I have, through the kindness of Mr. Rokujo, received a wild specimen of this species. It is totally distinct from *H. maritima*, being bone-hardy, with very slender red stems, narrow-sepalled white flowers turning red in sunlight and with long, narrow, matt leaves.

Section II CORNIDIA

In this large section of evergreen lianas found in Mexico, South America, the Philippine Islands, Formosa and, probably, other Pacific Islands, we have only *H. integerrima* in cultivation in this country as yet.

The other species of this section, which are not in cultivation, are only briefly described, from the horticultural angle, in the notes which follow. Full botanical descriptions in Latin, from herbarium material only, by J. Briquet may be found in *Annuaire du Jardin Botanique de Genève*, **20**: 179–202 [1916–1918] and brief diagnoses in German by the late Adolf Engler in *Die Natürlichen Pflanzenfamilien*, 2nd edition, **16a**: 207, etc. [1930]. Other sources are cited with the individual species.

Those who would prefer to skip the botanical data may proceed straight to page 83 where a horticultural summing up of this strange and fascinating section of the Hydrangeas is given.

The Section CORNIDIA is botanically described by Adolf Engler as follows:

Shrubs and trees or lianas whose young growths develop, in places, adventive roots like Ivy. Leaves evergreen, coriaceous (except in *H. Goudotii*). Petals of fertile flowers mostly free, but in *H. peruviana* hanging together in a cap. Ovary inferior. Capsule truncate. Seeds narrowly linear, curved, very small. Involucral bracts of inflorescence soon falling.

Subsection 1 MONOSEGIA

Inflorescence consisting of a single false umbel whose primary branches are provided, at the base, with involucral bracts. Styles finally thickening into a column.

Series SPECIOSÆ – having sterile flowers.

H. Seemanni Riley is a species from Mexico, found in the Sierra Madre in ravines 'climbing and rooting on old trees like Ivy' (Seemann, *Bot. Voy. Herald*, 293 [1856]). It is described in the *Kew Bulletin* (207 [1924]) as distinguished from *Oerstedii* by its flower-stalks being very short and the leaves being narrow, hairy beneath and yellowish-green on both surfaces. From Standley (*Trees and Shrubs of Mexico*, **23**: Pt. 1, 308 [1920]) we learn that the leaves are lanceolate or elliptic-lanceolate, narrowed at the base and up to 6 inches long and 2½ inches wide, yellowish-green, leathery, and conspicuously veined. The inflorescence is not downy, with primary flower-stalks only an inch long or less (instead of the usual 6 to 8 inches) and the flower clusters are laxly arranged in racemes. The specimens show nine or more fine ray-flowers, each 1½ inches or more across and having four entire, rounded sepals, apparently pink in colour. They project from the umbel on long and very slender upright stalks. The bud is enclosed in large rounded bracts before expansion in the usual manner. It is evidently a highly decorative species and why American plantsmen do not appear to have been able to secure the fine thing for the many beautiful gardens in the warmer zones of their country it is hard to understand.

H. asterolasia Diels, a species from Ecuador, is described as a climbing or epiphytic shrub up to 30 feet, with the young branches and undersides of the leaves red-furred. The leaves are narrowly obovate and acute at the apex, with entire margins and from 2–4 inches long and 1–1½ inches wide. The flower-stalks are also red-furred and the outermost flowers are sterile and over an inch across and yellowish-white in colour. The habitat is given as Mera, Pastaza, Eastern Ecuador, at

3,000 feet in remains of primæval rain-forest. The flowering time is August (*Notisblatt Bot. Gart. Berlin,* XV, 370 [1941]). The species appears to resemble *H. integerrima* in leaf more than the other members of this series.

H. epiphytica Morton is a remarkable climbing species, with small fluffy umbels of crowded fertile flowers on red stalks, with many tiny white ray-flowers. The leaves are very small for this section, being only 2 to 3 inches long and most often lanceolate, sometimes elliptic. It was found in Costa Rica at Vara Blanca on the north slope of the central Cordillera between the volcanoes of Poas and Barba at an altitude of 1,680 metres.

H. integrifolia Hayata is described as similar to *H. integra* but the leaves are more ovate and broader at the base; the stems are covered with dark red bark and the young shoots are densely woolly at first, becoming smooth later. Leaves ovate. The sterile flowers have two to three sepals. Found on rocks at 1,500 metres altitude at Taitou, Formosa.

H. integra Hayata is a Formosan Hydrangea of this section. In a letter to me, Mr. H. K. Airy-Shaw kindly sent the following notes on this species:

'In 1912 Mr. W. R. Price collected fruiting inflorescences on Mt. Arisan, climbing on Chamæcyparis, at 7,000 feet. There are at least 12 ray-flowers but colour not stated (probably white).

'Then, about 1930, Yashiroda got it in flower on July 30th. He says: "The *Hydrangea integra* was very grand plant. The specimen least render its grandeur and loveliness in flower and foliage."

'Yashiroda was formerly a student gardener at Kew and so could appreciate things from the Horticultural angle.'

Yashiroda's specimen appears to have small ray-flowers with sometimes three or four sepals, but it is not very clear. My only further information on this species is that Hayata recorded it as growing at elevations of 1,660 to 2,830 metres and described it as having only a few two-sepalled ray-flowers. (*Icones Plantarum Formosanarum,* **3:** 108 [1913].) In leaf, shoot and bract characteristics it follows the general run of this section.

H. glandulosa Elmer is a magnificent but very rare and local species from the Philippines. I am glad to say that not only was its discoverer one of those rare collectors whose field notes give us a real picture of the plant in the wild, but the Kew Herbarium is fortunate in having a perfect dried specimen beautifully preserved. This is all I have seen, but it shows a noble inflorescence 6 inches across and of the greatest beauty. There are as many as thirty orbicular, four-sepalled ray-flowers each over an inch across.

The following description is taken from *Leaflets on Philippine Botany*, **2**: 473 [1908] by A. D. E. Elmer.

'A powerful tree climber, stems about 1¼ inches thick. Branches quite rigid and characteristically curved and crooked. Leaves scattered, opposite, lucid, very rigid, dark green above, paler beneath, 4 inches long, 2½ inches wide, rounded, oblong. Inflorescence terminal, yellowish or rufous tomentose (woolly), corymbosely spreading, nearly 4 inches wide. Sterile showy flowers, petaloid calyx lobes rotate, 1¼ inches in diameter, segments four, nearly free, creamy-white.

'Type specimen from Dumaguete, Cuernos Mts., Province of Negros Oriental, Ap. 1908.

'A tall tree climber, forming tangled bushes and more or less rambling over tree-tops on a very steep wooded ravine at 3,000 feet, near the base of the main precipice of the highest peak on its eastern side. Only two or three plants seen in this almost inaccessible place. A very rigid species with glands on the underside of the leaves and *fragrant flowers* [my italics].

'Rare and local, mossy forest 900 to 2,200 metres, endemic.'

H. cuneatifolia Elmer, from the Philippines, is a lofty tree climber or, rather, epiphytic, that is to say, growing on other plants but not parasitic. The stems and branches form more or less rigid clumps, the free branches hanging. The flowering twigs are ascendingly curved. The leaves are leathery, with recurved tips, opposite, and scattered along the long fulvous stems. The margins are entire, obovately oblong, wedge-shaped at the base, smooth, and of a paler green on the undersurface. In size they are 6 inches long by about 2½ inches wide. The globose involucre of the buds is about 1½ inches in diameter. The bracts composing it are three or more in number, smooth and yellowish-white on the inside and densely brown-felted outside, succulent and falling off very early. The numerous outer or terminal flowers are sterile, an inch or more across and have three or four creamy-white sepals. The 'type' specimen comes from Todaya, Mount Apo, District of Davao, Mindanao.

Its habitat is described as 'dense forested flats' at 3,000 to 4,000 feet elevations. A clue to the reason for the strange fact that these superb plants have remained so long undiscovered and brought into cultivation is given in the following sentence from *Leaflets on Philippine Botany*, by A. D. E. Elmer, **7**: 2825 [1915]. 'The fallen bracts are frequently met with under trees, but the plant inhabits the uppermost limbs of lofty or giant trees clear out of sight of the travelling botanist.'

H. Oerstedii Briquet is a climbing species from the high mountain forests of Costa Rica and the mountains of Veragua, Panama. It was discovered by Oersted at elevations of about 6,000 feet. The ray-flowers surrounding the 10-inch-wide corymb are described by Briquet, from a dried specimen, as few, but each about 1½ inches across.

But further notes on *H. Oerstedii* are to be found, by Standley, in *The Flora of Costa Rica* (in *Publ. Field Mus. Nat. Hist.* **18**: 473 [1937]). He describes it as common in the forests of the central mountains at 1,800 to 3,000 metres elevations, also in Guanacaste at 1,200 to 1,300 metres and in Panama. Apparently it may be a climber reaching the tops of tall trees, or an erect shrub, or a tree. It has little or no down on leaves or shoots. The numerous outer flowers of the large cyme are sterile and an inch or more across and of a delightful pink colouring, making it one of the most beautiful plants of Costa Rica, its showy flowers being quite as decorative as those of the best cultivated Hydrangeas.

A most interesting point is that 'the young plants of this species, with small denticulate leaves, abound almost everywhere in the temperate region, growing on tree-trunks, and even fence-posts, but they are so unlike adult plants that no one would suspect the relationship'. Apparently it is only when the log finally crumbles into the mould that the best placed of the young plants can then start to grow to the adult state. Thus it would appear a very easy plant to collect in quantity and I think it is high time that it was imported and tried out in cultivation here, as it seems to be a vigorous and efficient species.

H. durifolia Briquet has the usual leathery, dark green, oblong-elliptic leaves and a flower-head 5 inches across with four-sepalled, ¾-inch wide, red ray-flowers crowded around the margin. The colouring is described as intense. The habitat is given as La Baja in the Province of Pamplona, Colombia, at an elevation of about 7,000 feet.

H. Schlimii Briquet. The habitat is given as Colombia, in forests at 2,600 metres. It is described as a shrub having leathery, sub-ovate to elliptic leaves and dark red ray-flowers about ¾ inch across projecting on long pedicels in May.

H. Goudotii Briquet is a tree of 20 to 25 feet, with smooth leaves less leathery and hard than those of allied species. The flower-head is described as about 10 inches across with ray-flowers about 1½ inches across; 'flowering in January on the banks of the Combayma River, or, in June, at Ibaque and in the thickets of Quindiu at Portoihuelo, Colombia'. We are not told the colour of these flowers.

H. platyphylla Briquet. A noble Colombian species, is a climber from high mountain forests in the province of Mariqueta described by the collector, Jean Jacques Linden, as 'bracted with vivid red, flowers deep violet, in January'. The herbarium material and data indicate large leaves 5 to 6 inches long and an inflorescence about 10 inches across with numerous ray-flowers each over 1 inch across. Eminently desirable.

H. peruviana Séringe. A species from Ecuador with small, leathery, oval-elliptic leaves. There are only a few sterile flowers, about 1 inch across with rounded sepals, netted and veined. In colour they are, apparently, white.

H. Trianæ Briquet, from Colombia, at elevations of about 5,000 to 7,000 feet, is described as a rather similar species with smaller and more dense inflorescences with smaller, shorter-stalked ray-flowers. On the other hand, the specimens in the Kew Herbarium would indicate that this is one of the finest flowers of the whole section. The ray-flowers are large and numerous and are described in detail in my summing-up of this group at the end of this chapter. The locality given was the Province of Antioquia, Colombia, at an altitude of 2,600 metres and the flower colour is described as intense brownish-red.

H. panamensis Standley has curiously shaped smooth, obovate leaves notched at the ends, with a flower-head only 2½ inches across but having six to nine tiny ray-flowers. It opens, as usual, from a bracted bud.

In *New Plants from Central America* (*Journ. Wash. Acad. Sci.*, **17:** 10) Standley describes the fertile flowers as yellow with purple calyces and the bracts as purple. It is interesting as being the only lowland species of this section so far recorded. The specimen came from 'along Rio Fato, Colon, Panama', at only from 10 to 100 metres altitude.

H. Weberbaueri Engler is a Peruvian species having sterile ray-flowers with red petaloid sepals and the usual climbing habit and large, ovate, leaves.

H. caucana Engler has an inflorescence 3 inches across with, apparently, five or six dark red ray-flowers. Found at elevations of 4,000 to 5,000 feet, in Colombia.

H. Lehmannii Engler, a Colombian species, is described as near *H. Trianæ*, but frequenting higher altitudes (2,000 to 2,400 metres) and having oblong instead of ovate leaves and these are narrowed at both ends.

H. Preslii Of little horticultural interest, having no sterile flowers.

H. Briquetii (*H. umbellata*) from Perú. Again of little horticultural interest.

H. ecuadorensis Briquet. Of little horticultural interest.

H. diplostemona (Donn. Sm.) Standl. comes from Costa Rica (see *New Plants from Central America* in *Journal Washington Acad. Sci.*, **7:** 9) where it has been found near el General, Province of San José, at an altitude of 1,070 metres. It is described as a forest liana with no sterile ray-flowers, but the numerous fertile flowers pink or red. The large, thin leaves are more pronouncedly acuminate and the young shoots are covered with reddish down at first. The type was collected 'in forests at Tais, Atlantic slope of Costa Rica at an altitude of 700 metres', in 1900.

H. Sprucei Of little horticultural interest. Said to have 'wine-red flowers'.

H. Steyermarkii Standley a Guatemalan species found growing on tree-trunks, with branches covered with dense, thick brown scales and loosely woolly brownish hairs, and having minutely toothed, leathery leaves. Inflorescence axillary, ¾ inch high and 3 inches broad. Flower-stalks covered with brown hairs, bract ¾ inch long and broad. No ray-flowers. See Standley, *Pub. Field. Mus. Nat. Hist. Chicago Bot.* Ser. XXII, 233 [1940].

Subsection POLYSEGIA

Inflorescence consisting of several false umbels placed one above the other. Primary flower-stalks in young stage enclosed by several bracts. The foregoing is Adolf Engler's description of this subsection but, after examining a wide selection of herbarium material, I am uncertain of his placing of some of the species. It appears to me that in some cases it is merely a question of a more free-flowering nature, otherwise the arrangement is much the same as that of many members of the Monosegia. On the other hand, some species do have an actual flower-stalk which branches into three, so that two subsidiary umbels appear below the top one, sometimes giving the strange flower the form of a decorated cross. Adolf Engler's specimens were destroyed in the Berlin Herbarium during the late war and the whole section therefore really requires working out again. But the species placed in this subsection are characterised by having aborted, misshapen, or otherwise unsatisfactory ray-flowers.

Another species is **H. serratifolia** (Hooker et Arn.) Engler from the Island of Chiloe, with smooth, serrated leaves and curious additional husk-like bracts on the lower part of the flower-stalks.

H. integerrima (Hooker et Arn.) Engler. This evergreen climbing species comes from Chile and though long known, being described by Séringe in De Candolle's *Prodromus* in 1830, was only introduced by H. F. Comber about 1927.

Climbing, up to 100 feet, up tall trees and on rocks, assisted by aerial roots, it has entire, thick, leathery, vivid dark green, elliptic, pointed leaves. The unusually abundant flower-heads open from globular buds enclosed in bracts and, so numerous are the inflorescences on short lateral stems that a branch has the appearance of a panicle composed of numerous small corymbs of the little white fertile flowers. Each corymb, in the form available here, has only one irregularly formed sterile white ray-flower about an inch across. More often than not, the ray-flower consists of only a single, whitish sepal, stained with green and of variable size and shape. Although this form of *H. integerrima* has no great flower beauty, it is a useful addition to the very few evergreen, self-clinging climbers in cultivation. In the wild the trunk is sometimes 4 inches in diameter.

I find it difficult to follow the dictum of changing the name *integerrima* (which means entire or unserrated) to *serratifolia* (which means having serrated leaves). An enormous plant covers the north wall, a chimney stack and part of the roof of my house, and not one of its leaves is serrated. It is, indeed, the only Hydrangea that I have seen that does not have serrated leaves! They are entire and smooth without hairs.

Evergreen, self-clinging flowering climbers are not common – in fact I can only call to mind *Bignonia capreolata*, *Trachelospermum asiaticum*, *Mitraria coccinea* and *Berberidopsis corallina*, and these need a trellis for real stability. Thus I think *H. integerrima* may well become quite a popular climber for walls.

A specimen of this species in the herbarium at Kew has shapely four-sepalled ray-flowers, each over an inch across and there are four or five of these to each inflorescence. Thus it would appear that much better forms of this species than the one which is available in commerce here may exist.

H. subintegra Merrill, a species from Bataan, Philippines, resembles *H. integra* Hayata. The leaves are smaller and the sterile flowers have four sepals instead of two, a most important point. It is also distinct by reason of its erect but not climbing habit. The following description is taken from the *Philippine Journal of Science,* **3** [1908], Botanical Section, p. 408:

'Shrub 5 feet high, erect, glabrous except inflorescence. Branches terete, reddish-brown, smooth, shining. Leaves opposite, oblong lanceolate, membranaceous, 3 to 4½ inches by ¾ to 1¼ inches, upper surface brownish when dry, lower side paler, shining. Margins entire, sometimes denticulate distantly and obscurely. Apex acuminate, base acute. Cymes terminal, 2 to 3 inches long, slightly

Hydrangea paniculata floribunda

Hydrangea paniculata

X hydrangea macrophylla 'Générale Vicomtesse de Vibraye'

fulvous-hirsute. Outer flowers sterile with four sepals which are petaloid, obovoid to elliptical-ovoid and about half an inch long, glabrous and white in colour. One sepal is usually larger than the other three. Distinct from *H. Lobbii.*'

The foregoing is another of the rare cases in which the flowers are described in such a manner that we can visualise their appearance. (Helen McClintock has classified this under *Hydrangea Scandens* Poeppig.)

H. tarapotensis Briquet is of little horticultural interest having no showy sterile flowers.

H. Bangii Engler and *H. antioquiensis* Engler from Colombia have the fertile flowers white. As they are without the well-formed, showy ray-flowers which are the great beauty of the Hydrangeas and the principal reason for their cultivation, I omit descriptions here and will refer the reader to Engler and to Briquet as cited at the head of this chapter.

H. Mathewsii, with rufous-felted leaves and branching flower-stalks, is of little horticultural interest.

My descriptions of these strange evergreen Hydrangeas of the Cornidia section are mostly based on dried specimens and botanical descriptions in Latin or other languages which in many cases devoted much space to exact data regarding the down covering this or that part, the form of the veins, etc., etc., but omitted the colour, size, shape and numbers of the sterile flowers. They were silent on this most important point, from the horticultural angle.

Few Europeans have seen these curious flowers alive and the field notes of the collectors are often meagre. From a perusal of all the information that I could find and from examination of the herbarium specimens available I have, for what it is worth, built up the following picture of these extraordinary Hydrangeas in my mind.

The species are distinct in character, but evidently very closely related and follow a recognisable pattern. Evergreen scramblers for the most part, their wood has a vine-like appearance. The evergreen leaves are large, on the whole, about 6 inches long being a common size. Their shape varies from lanceolate to ovate and obovate. The man in the street would probably call them rather Rhododendron-like. The unusual feature of these plants is the peculiar form of the flower-head. It begins as a large flower bud, an inch or more wide at the base, on an otherwise bare stalk. One would imagine that a Magnolia-like flower was about to expand. The stalk is unusual in that it does not taper, it is of even thickness right up to the base of the bud. When this bud opens, it forms a flower-like chalice of bract-like, rather than petal-like, material. Sometimes this chalice is coloured a vivid red. Otherwise, one is rather reminded of the bracted flower of – say – *Cornus Nuttallii*. But, instead of the central pompon of that inflorescence, a typical Hydrangea flower-head rather resembling that of a Lacecap variety of *H. macrophylla* erupts from the centre of the chalice of bracts which then falls away.

To give an example; in the herbarium specimen of *H. Trianæ* at Kew, this flower-head, umbellate rather than corymbose, carries apparently from eight to eleven typical, large Hydrangea ray-flowers which, in their dried state, I could not have distinguished from those of *H. a.* var. 'Bluebird'. Only vestiges of the frilled, cup-like, basal bract remained on the specimen.

As to that vital point, the colour of these fine ray-flowers, described as intense red as they often are and even showing that pigment in their dried state, I have an uneasy feeling that possibly they may open a greenish-white and only attain that red – like *H. s.* 'Grayswood' or *H. m.* 'Hamburg' – as they age.

This is only a suspicion without any definite foundation. If any of these species do, indeed, open red flowers, they should be secured at once. *H. integerrima* is reasonably hardy and its best forms (seen only as dried specimens) are evidently quite as decorative as *H. petiolaris*. From these materials the hybridist should be able to produce a red-flowered, evergreen, self-clinging climber hardy enough for fairly general use on shady walls. Such a plant would obviously have an immediate and enduring commercial success. Every garden has, at least, house walls and, due to the dearth of self-clinging evergreen flowering climbers, all sorts of non-climbing substitutes such as Cotoneasters and Pyracanthas have to be used with unsatisfactory results.

The habitat of the Cornidias is generally mossy forest with a high rainfall and a humid atmosphere, their homes are in the last few beautiful wild places in the world, and Richard Spruce stated in his *Amazonian Journal* that he had never seen a Cornidia in either the hot or the cold zone in South America, they always grew in the merely warm zone.

THE CULTIVATION OF HYDRANGEA SPECIES

In general, a fairly moist rich soil is preferred and light shade is desirable. Many will grow in a naturally chalky soil, but to achieve fine plants under these conditions the addition of humus and the provision of some shade are necessary. For the most part, the Chinese species are quite happy in alkaline soils, whereas the Japanese are less so.

Protection for the young growth from slugs is essential for young specimens of most of the species. No pruning is needed. Indeed the reason for the strange fact that many species that are no hardier than × *H. macrophylla* are found as strong bushes in gardens where that plant is reduced to the status of a non-flowering herbaceous wreck, is that the species have not been pruned and *H. macrophylla* has been so treated at the wrong time of year and then left to its fate.

On all other points, culture and propagation follow the lines indicated for × *H. macrophylla*.

The Cornidia section evidently require protection from sun-heat and dryness at the roots. Although they inhabit warm regions, the behaviour of *H. integerrima* indicates that at least some species will prove good garden plants. Hybrids, when they appear, should be still more amenable.

On the whole, cultural conditions which suit Rhodo-dendrons may be expected to suit the Hydrangea species. By planting them in association, the woodland Rhododendron garden may be made as attractive in late summer as it was in the earlier part of the season and the gay and vivid hues of Hydrangea foliage will be found to provide a pleasing contrast to the more sombre tones of the evergreens when the latter are in flower.

Summing up the cultural side, it may be said that the AMERICANÆ are easily satisfied and mostly hardy and good garden plants. The PETALANTHE are highly decorative and responsive to good cultivation, preferring a rich soil and light shade, with the exception of *H. maritima*, which likes the seaside and full exposure. The HETEROMALLÆ are hardy and robust for the most part and not very particular. The ASPERÆ are rather more fastidious as garden plants, but highly reward-ing when success is attained. Most insist upon a situation that is not either sunbaked or dank, a soil that is never sodden or dried out and an atmosphere that is pure and neither too dry nor too wet. The members of the CALYPTRANTHE subsection are easily pleased. Those of the CORNIDIA section have, I think, the typical requirements of South American plants: a moist atmosphere and a mild climate. I should imagine that they would succeed well wherever *Desfontainea spinosa* and *Tricus-pidaria lanceolata* flourish.

The greatest difficulty will be in obtaining the plants, but the British people have a way of getting what they want, and no gardeners in the world are better served by an efficient and active horticultural society, with its affiliated organisations, and a progressive and well organised horticultural industry.

Chapter 3

Hydrangea Macrophylla garden varieties

The pure wild species from which this hybrid race has been evolved are dealt with in the section devoted to these. This chapter deals with the garden-bred hybrids. For convenience we may divide these into two sections – the Hortensias, bred primarily for pot-work and characterised by a globose corymb chiefly composed of sterile flowers, and the Lacecaps, bred for use as flowering shrubs, whose flower-head is more or less flat or disc-shaped, having the centre part filled with the tiny fertile flowers and an outer ring of the relatively very large, sterile, ray-flowers.

Taking the Hortensias first, these are mostly the products of French, German, Swiss or Dutch breeders. The variation in the five-hundred-odd named varieties, taking outdoor plants for comparison purposes, is astonishingly slight. But this is because they have been bred with a fixed pattern as the target. A huge globose-head was the sole aim. And no outcross with a new species was used.

The individual flowers vary in size from a width of eight centimetres to three centimetres, in colour from pale pink or pale blue through lilac, purple or violet to crimson or deep blue and, much more rarely, to white. In shape the individual flowers may have the perfectly formed four-sepalled, rounded shape of 'Westfalen', each standing out separately from its neighbour, or they may be ragged, with pointed sepals, and produced, in different sizes, in such a dense mass that one cannot see any one flower clearly; or, again, the head may be composed of a smaller number of large flowers with four or five large sepals having beautifully serrated edges, like 'Altona' or 'Hamburg'.

In height the Hortensias range from a dwarf bush 12 inches by 18 inches, such as 'Vulcain' or 'Vendôme', to an 8-foot mound such as 'La Marne', 'Mme E. Mouillère' or 'Joseph Banks'.

I have drawn up the following lists from my own experience and observations in growing the plants outdoors, also from many valuable data sent me by Mr. Jessup, Director of the National Herbarium and Botanic Gardens, Melbourne, where the Hydrangea is the most popular shrub for Christmas decoration, from Marcel Ebel's admirable little French book *Hydrangea et Hortensia,* and from information supplied by Monsieur Mouillère, Monsieur Cayeux, Monsieur Lemoine, Monsieur Kluis, Herr Wintergalen, Herr Matthes, Prof. Dr. H. J. Venema of Landbouwhogeschool, Wageningen, Mr. G. J. Baardse, Herr H. Schadendorff, Mr. J. C. Wezelenburg, Monsieur Draps, Herr Moll, Herr Otto Armbrecht and other breeders and growers and by gardening friends. The remarks in this chapter on the comparative value of the varieties and the colours of the flowers are based on their performance as outdoor flowering shrubs in the garden. The varieties recommended for pot-work are listed and discussed in another chapter.

In the hope that it may assist the amateur gardener requiring only a few plants for his garden to make a quick selection I have ventured to accord a star of merit indicating proved excellence as a garden shrub. Certain outstanding varieties which are notably effective for the main plantings are accorded two stars. When a variety is not well known to me personally I have limited the particulars to the brief ones given by one or other of the authorities cited above. In the colour descriptions I have used the colour names shown in the Royal Horticultural Society's *Horticultural Colour Chart*★ wherever possible. Without this exact evaluator, colour descriptions are apt to be misleading and inaccurate. The colours described are those borne by open-ground plants outdoors. Many varieties which do not readily give blue flowers outdoors can be made to do so quite easily when grown in pots, by suitable feeding.

In the description of the growth I have cited as dwarf those attaining little over a foot in height with good cultivation in

★ First edition.

average soil. Moderate growth signifies a plant attaining about 18 inches to 2½ feet. Those described as moderate to tall are about 3 feet. The varieties described as tall are above this height. All these Hydrangeas can, of course, build up to much greater heights in favourable gardens by extending further and further on old wood. I have therefore taken as my guide the average length made by a strong shoot from the base of an established plant during one growing season.

THE HORTENSIAS

Variety	Breeder	Remarks
AALSMEER'S GLORY	D. Baardse, 1930	Deep pink or mid-blue, moderate vigour.
ADELAIDE	H. J. Jones, 1927	Rich pink, large spreading corymb, moderate growth.
ADMIRATION	H. Cayeux, 1932	Compact heads, self-coloured crimson or deep blue.
AGNÈS BARILLET	Mouillère, 1909	Creamy-white, large head.
ALPENGLUHN		Crimson, mid-season.
ALTONA★★ ('ALTHONE')	H. Schadendorff, 1931	Moderately tall, stout growth, large head, serrated sepals, Rose Madder or, with feeding or in acid soil, vivid deep blue; lasts well, the heads turn later to attractive autumn flower tints; mid-season flowering, requires shade. Award of Merit 1957.
AMARANTHE	Mouillère, 1920	Large waved flowers, amaranth pink, moderate vigour, short stout growth.
AMAZONE	Lemoine, 1918	Pure white, serrated sepals.

Variety	Breeder	Remarks
AMERICA	Mouillère, 1928	Deep pink or blue, large corymbs, late.
AMETHYST★	M. Haworth-Booth, 1938	A seedling of 'Europa' with double, serrated-sepalled, pink or amethyst-shot Flax Blue (H.C.C. 642/1) flowers in small heads produced with unusual freedom from August until frost; at its best in early September. Shoots stout but of moderate height, leaves very thick and deep green. Not suitable for pot-work, and not a very ready 'bluer' until matured. Illus. 20.
AMI PASQUIER★	E. Mouillère, 1935	Dwarf to moderate growth, a superb crimson variety, continuous flowering. In acid soil wine-purple or blue, colour does not fade. A seedling of 'Maréchal Foch'. A.M. 1953.
APOLLO	H. Cayeux, 1932	Bright crimson-pink or deep blue flowers, good autumn tints and a ready 'bluer'. Mid-season, moderate to tall growth.
ARCHIE MOWBRAY	H. J. Jones, 1927	Rosy-mauve.
ARMAND DRAPS		Pink. Early.
ARTHUR BILLARD★	E. Mouillère, 1935	A compact, moderate to dwarf variety, deep crimson or deep blue or violet flowers; mid-season flowering.
ATLANTIC		Red. Mid-season.

20 × *Hydrangea macrophylla* 'Amethyst'

Variety	Breeder	Remarks
ATTRACTION	Foucard, 1912	Pale pink, strong wood, free-flowering.
AUKMANN	Moll, Zürich	Large vivid pink corymbs or dark blue in acid soil. Dwarf growth, well thought of in Australia.
AVALANCHE	Lemoine, 1908	White, weak branches.
AYESHA	Imported	Lilac-like heads. Unusual.
BAARDSE'S FAVOURITE	D. Baardse, Aalsmeer, 1920	In neutral soil, warm pink, mid-season, moderate growth.
BABY BIMBENET	Mouillère, 1910	Pink or pale blue, free-flowering, short stout growth. Does not last well.
BAGATELLE	D. Baardse, 1922	Rose pink, mid-season to late flowering, lilac-blue in acid soil, moderate but stout growth.
BARONESS SCHRÖDER	K. Wezelenburg, 1927	Flesh pink or pale blue, burns easily in sun and is tender. Large corymbs.
BEAUTÉ HAVRAISE	H. Cayeux, 1919	Warm pale rose, serrated sepals, rather leggy.
BEAUTÉ ROSE	H. Cayeux, 1930	Seedling of 'Normandie', strong stems, compact corymbs, large individual flowers of clear rose or, in acid soil, the colour of Parma violets.
BEAUTÉ VENDÔMOISE*	Mouillère, 1910	White flushed pink, moderate growth, many fertile flowers, the sterile ones very large, 4½ inches across, see Lacecap section.
BEAUTY OF DRESDEN (SCHÖNE DRESDNERIN)	F. Matthes	Pink ripening to claret colour, free-flowering, short strong growth.

Variety	Breeder	Remarks
BELGICA		Deep pink or mid-blue, vigorous growth and good constitution.
BENELUX		Stout growth.
BLAUER PRINZ★ (BLUE PRINCE)	F. Matthes	Pink or purplish mid-blue if fed aluminium, pointed sepals, overcrowded corymb, not lasting but opening successively and therefore effective. Mid-season to late-flowering, moderate height.
BODENSEE		Readily blue.
BORDEAUX	H. Cayeux, 1938	Flowers the colour of claret.
BOUQUET ROSE	Lemoine, 1908	Pink or clear blue, vigorous and free-flowering weak stems.
'BRIGHTNESS' (see GLOIRE DE VENDÔME)		
CANDEUR	Lemoine, 1921	Pure white flowers with serrated sepals, large corymbs.
CARL SPITTELER	Gebr. Moll, 1946	Deep rose, large corymbs, free-flowering, mid-season.
CARMEN★	E. Mouillère, 1936	A dwarf, upright plant that requires care, vivid warm crimson or purple flowers with serrated sepals, very beautiful where it will grow well. Good in Australia. Mid-season flowering.
CARMEN★	Wintergalen, 1938	Coppery crimson, mid-season, vigorous.

Variety	Breeder	Remarks
CAROLINE		Bright pink or bright blue flowers, tall growth, large flower-head lasting well on the bush. Good in Australia.
CAROUSEL		Red, mid-season.
CENDRILLON	Lemoine, 1921	Deep pink or blue, dwarf growth.
C. F. MEYER	Gebr. Moll, 1946	Pink or blue, large individual flowers, clear colour, dwarf growth, dark green foliage, free-flowering, late.
CHARME	H. Cayeux, 1932	Large flowers self-coloured deep pink (Rosebelle × Normandie).
CHARMING	Emil Draps, Strombeck, 1938	Pink or Gentian blue, mid-season.
COCARDE	H. Cayeux, 1932	Long-lasting, large, round corymbs, reddish-pink
COLONEL DURHAM	K. Wezelenburg, 1927	Clear pink or pale blue, large corymbs but does not last well.
COLONEL LINDBERGH	D. Baardse, Aalsmeer, 1927	Pink or lilac, mid-season to late, moderate growth.
COQUELICOT	Mouillère, 1922	Bright crimson, early and free-flowering, stiff growth.
CORSAIRE	L. Cayeux, 1937	Large self-crimson flowers.
COVENT GARDEN	K. Wezelenburg, 1937	Sturdy, large leaves, medium height, large serrated sepals, pink or, readily, mid-blue. Mid-season flowering.
DAPHNE	J. Wintergalen, 1937	Crimson, short strong growth, mid-season. In acid soil a port wine colouring is achieved.

Variety	Breeder	Remarks
DARLING (see MEIN LIEBLING)		
DAVID INGAMELLS	H. J. Jones, 1927	Warm pink, moderate growth, dwarf compact habit. A.M. 1928.
D. B. CRANE	K. Wezelenburg, 1927	Pink, large flowers slightly serrated, dwarf growth. A.M. 1927.
DENTELLE	Lemoine, 1909	Pale creamy-pink, serrated sepals, dwarf growth.
DEUTSCHLAND★	D. Baardse, 1921	Deep Rose Madder or purplish blue if fed aluminium, large flowers, foliage attractively tinted in autumn, mid-season to late. A.M. 1927.
DIANE	Lemoine, 1910	Palest pink, large corymbs of crimped flowers.
DIRECTEUR VUILLERMET	Mouillère, 1913	Pale pink, large corymbs.
DÔME FLEURI	Lemoine, 1911	Pale pink, bun-shaped heads.
DOMOTOI★	An old Japanese variety	Individual flowers large and doubled, pale pink or blue; an attractive irregular head for a change. Growth vigorous but not tall. Apt to become fasciated and distorted if too well fed. Mid-season to late.
DRAPS' PINK	Emil Draps, Strombeck, 1938	Rose or Solferino purple.
DUCHESS OF YORK	H. J. Jones, 1927	Rose pink, dwarf habit.
EARLY RED		Red, early.
EASTER STAR		Deep pink, mid-season.
ECLAIREUR	Lemoine, 1913	Vivid crimson-pink, dwarf growth, free-flowering.

Variety	Breeder	Remarks
EDELWEISS	P. Flores, Wuppertal-Sonnenborn, 1931	Creamy or greenish-white, moderate growth.
EDISON	K. Wezelenburg, 1936	Pink fringed sepals, sturdy growth.
E. G. HILL	Lemoine, 1912	Pink or blue, very free-flowering, a first-cross between *rosea* and 'Otaksa'.
ELBE	H. Schadendorff, 1929	Mid-season, pale pink or faint blue, dwarf growth.
ELDORADO		Deep pink.
ELECTRA	H. Cayeux, 1932	Pink, large corymbs.
ELÉGANCE	L. Cayeux, 1937	Small compact heads, deep purplish-pink or intense blue.
ELITE		Deep pink, mid-season.
ELMAR	J. Wintergalen, 1928	Pink or purplish, serrated sepals, stout growth, does not give blue flowers easily. A.M. 1924.
EMBLÊME	H. Cayeux, 1926	Crimson flowers but a weak grower.
EMILE MOUILLÈRE	Mouillère	White, blue eye, moderate growth.
EMOTION	L. Cayeux, 1937	Deep pink, vigorous, solid corymbs.
ENZIANDOM★★		Very good deep blue.
ETINCELANT	Lemoine, 1915	Pink flowers, moderate to dwarf growth, free-flowering, mid-season. A.M.
EUGÉNIE TABART	Mouillère, 1909	Pale pink.
EUROPA★★	H. Schadendorff, 1931	Pink or vivid pale blue, tall and very vigorous, mid-season to late, large heads.
EXCELSIOR	Chaubert, 1922	Deep pink, large heads.
EXTASE	L. Cayeux, 1938	Fresh deep pink, well held.

Variety	Breeder	Remarks
FANAL ('FAVOURITE')	H. Cayeux, 1935	Large crimson flowers, dwarf growth.
FELIX	Gebr. Moll, 1946	Creamy-white, large corymbs free-flowering, glossy leaves, mid-season.
FISHER'S SILVERBLUE★★	Moll, Zürich	Pink or, if fed, blue, shapely flowers, very early, dwarf growth, very free-flowering. Colour fades in sun.
FLAMBARD	H. Cayeux, 1929	Compact heads, pink turning red in sun, dwarf growth.
FLAMBEAU		Vivid red.
FLAMINGO		Light red, early.
FLORA GAND		Deep pink.
FLORALIA		Pink, very early.
FLORENCE BOLT	K. Wezelenburg, 1927	Soft pink or pale blue, large corymb, moderate growth.
FLORISSE	Lemoine, 1911	Pale pink, creamy-centred florets.
FOCH (see MARÉCHAL FOCH)		
FORTSCHRITT	F. Matthes, 1931	Cherry pink or port wine to purple, moderate growth, early. Good in Australia.
FRAICHEUR		White, blushed pink.
FRANS HALS	K. Wezelenburg, 1936	Pink, dwarf and free-flowering.
FREYA	Wintergalen, 1928	Deep Rose Madder, serrated sepals, similar to 'Parsifal', purple in acid soils.
FRIEDRICH MATTHES	F. Matthes	Warm pink, dwarf but stout growth.
FRÜHLINGSER-WACHEN	H. Schadendorff	Deep pink, early.

Variety	Breeder	Remarks
FULGURANT	L. Cayeux, 1937	Stout growth, large compact heads, deep pink or purple flowers.
GALATHÉE	Lemoine, 1911	Pure white, serrated sepals.
GARTEN-BAUDIREKTOR KUHNERT	F. Matthes	Pink or vivid French Blue (H.C.C. 43/1), mid-season; moderate to strong growth, not very free-flowering.
GÉNÉRALE VICOMTESSE DE VIBRAYE★★	Mouillère, 1909	One of the purest among the lighter blues (H.C.C. 645/1), tall slender growth, large overcrowded heads on long stems, early to mid-season, very free-flowering, but not successional. Bred from 'Otaksa' and *rosea*; requires light shade. A.M. 1947. (See Plate C.)
GENERAL GEKE		Deep pink, tall growth.
GERDA STEINIGER		Bright pink, early.
GERMAINE MOUILLÈRE	Mouillère, 1920	Flowers greenish-white, turning pinkish, serrated sepals.
GERTRUDE GLAHN★	F. Matthes	Deep pink, purple, rarely blue, moderate growth, nice firm heads, a healthy grower, mid-season to late, lasts well.
GISELHER	J. Wintergalen	Pink, early, weak growth.
GLOIRE DE BOISSY-ST.-LEGER	Nonin	Pink.
GLOIRE DE VEN-DÔME (syn. 'BRIGHTNESS')	L. Mouillère, 1935	Deep crimson, late-flowering and very long-lasting, dwarf growth but sturdy habit. Not very free-flowering.

Variety	Breeder	Remarks
GLORIA	H. Cayeux, 1930	Pink or mauve, dwarf growth.
GOLIATH★		Deep pink or purplish-blue, large flowers, small corymbs, vigorous and tall, mid-season. Not very free-flowering at first but superb by the sea.
GOTTFRIED KELLER	Gebr. Moll, 1946	Clear, bright rose-red, large corymbs, free-flowering, dwarf growth, mid-season.
GRACIEUSE	Lemoine, 1920	Pink, serrated sepals.
GRAF ZEPPELIN	F. Matthes	Deep pink, serrated sepals, strong compact growth.
GUDRUN	J. Wintergalen, 1923	Bright pink or blue, moderate, stout growth.
GYPSY		Red, early.
HAMBURG★★	H. Schadendorff, 1931	Massive growth and fine foliage; flowers mid-season to late, deep pink, purplish or deep blue, huge, boldly serrated sepals; lasts unusually long as the flowers show attractive autumn red colour probably derived from *H. japonica*. Requires shade to grow well and is only recommended for favourable gardens. (Plate D.)
HARMONIE	Lemoine, 1911	Blushed white, large corymbs.
HARRY'S PINK TOPPER		Salmon pink.

Variety	Breeder	Remarks
HATFIELD ROSE★	Stuart Low	A sport of 'Westfalen' and pure crimson, often with a white eye, moderate growth, a very beautiful Hydrangea, unusually hardy for a red and remarkably free-flowering. Mid-season. A.M. 1938.
H. B. MAY	H. J. Jones, 1927	Pink or blue, moderate growth. Large corymbs, compact habit. A.M. 1927.
H. J. JONES	H. J. Jones, 1927	Deep pink, dwarf growth, flat flowers, with rounded edges. A.M. 1928.
HEIDERÖSEL		Deep pink or blue, dwarf growth.
HEINRICH LAMBERT	Anton Rosenkränzer	Stout dwarf growth, vivid rose.
HEINRICH SEIDEL★★	F. Matthes	Fringed cherry-crimson flowers in a slightly overcrowded head, growth stiff and moderate to tall; useful as a red in favourable gardens, purple in acid soil. Vigorous growth but not free-flowering until matured. Mid-season.
HELGE	J. Wintergalen, 1921	Deep cherry pink or blue, early, moderate growth.
HIGHLAND GLORY	H. J. Jones, 1927	Crimson, tall.
HOLLANDIA	D. Baardse	Stout growth, large, toothed leaves, pink.
HOLSTEIN★	H. Schadendorff, 1928	Pink, or sky blue in acid soil, moderate, slender growth, mid-season flowering; shapely individual flowers with

Variety	Breeder	Remarks
		serrated sepals. Very free-flowering but needs shade.
HORNLY		Tiny dwarf, seldom flowers.
HORTULANUS WITTE	D. Baardse, 1915	Pink or blue, moderate growth, early flowering.
HORTUS	F. Matthes	Deep pink or purple, moderate, compact growth, free-flowering, mid-season.
IMMACULATA		White, dwarf, early.
IMPÉRATRICE EUGÉNIE★★	See p. 136,	A globose-headed hybrid of *H. japonica* with slender branches and pink flowers turning crimson in sunlight.
INNOCENCE	Lemoine, 1910	Creamy-white, serrated sepals.
ISLETTES (LES ISLETTES)	E. Mouillère, 1909	Pink, lasting well.
JAN STEEN	K. Wezelenburg, 1936	Pink.
JEANNE D'ARC	Jacket, Orléans, 1896	Pure white, slender but vigorous growth, pointed sepalled flowers. A sport of 'Thomas Hogg', with black stems.
J. F. MCLEOD	H. J. Jones, 1927	Pink or blue, compact heads, moderate growth. A.M. 1927.
JOCONDE	Lemoine, 1912	White, serrated sepals.
JOHN C. MENSING	D. Baardse, 1918	Deep pink, dwarf growth.
JOHN VAN DEN BERG	D. Baardse, 1915	Pale pink, tall grower.
JONKHEER VAN TETZ	D. Baardse, 1924	Warm crimson-pink, midget, discarded by many as it will scarcely grow at all.
JOSEF WINTERGALEN	Wintergalen, 1943	Warm pink.

Variety	Breeder	Remarks
JOSEPH BANKS ('HORTENSIS')		The original Chinese clonal variety introduced 1789, with shiny elliptic leaves, very stout coarse growth, making a huge bush with strong branches, 8 feet high. Corymbs very large, bun-shaped. Sepals open green and then turn pale pink or pale hyacinth blue. Some sepals carry a dark blue streak, as though made with a pen. A globose-headed sport of the pure wild maritime species, not a hybrid like all others. It usually flowers only from the terminal bud. Tolerates lime.
JOSEPH ISRAELS	K. Wezelenburg, 1936	Pink.
JUBILEE	H. J. Jones, 1927	Fuchsine pink, stout, dwarf growth.
JUPITER	K. Wezelenburg, 1936	Pink.
KING GEORGE	H. J. Jones, 1927	Rosy-pink, large sepals with serrated edges. A.M. 1927.
KING GEORGE	Emil Draps, 1938	('Rosebelle' × 'Merveille'.) Cherry pink or deep blue, moderate to tall growth, shapely flowers, late flowering, rather delicate when young. The stems are pale green and not spotted.
KLUIS SUPERBA★★	F. K. Kluis, 1932	('Foch' × 'La Marne'.) Deep pink, violet or deep

Variety	Breeder	Remarks
		blue, stouter than 'Foch', but colour fades, mid-season. Free-flowering.
KLUIS SUPERIOR	F. M. Kluis, 1945	('Kluis Superba' × 'Deutschland'.) Stiff, stout growth crimson-pink or mauve-purple.
KÖNIGIN WILHELMINA★	D. Baardse, 1922	Salmon-pink, mid-season, stout but moderately tall growth. Unusual colour.
KRIMHILD	Wintergalen, 1924	Fringed sepals, pink, dwarf to moderate growth.
LA FAYETTE	E. Mouillère et fils, 1932	Cherry-pink, moderate to dwarf growth, free-flowering, mid-season.
LA FRANCE	E. Mouillère, 1913	White with a coloured eye, serrated sepals.
LA FRANCE	D. Baardse, 1915	Phlox pink ('Cake pink'), or mid-blue, massive heads, slightly serrated sepals, tall growth, not free-flowering until matured. Mid-season.
LAKMÉ	Lemoine, 1920	Creamy-white turning pink, serrated sepals.
LA MARNE★	Mouillère, 1920	A tall, massive grower and late-flowering, pale pink or blue, very large heads with serrated sepals but not lasting. Superb by the sea.
LANCELOT	Wintergalen, 1920	Pink or blue, slightly serrated sepals, short compact growth.
LA PERLE	Mouillère	White, serrated sepals.
LE CYGNE	H. Cayeux, 1919	White, tall grower.
LE LOIR	Mouillère, 1909	Pale flesh colour, large corymbs.

Variety	Breeder	Remarks
LE LOIRET	Foucard, 1913	Pink or purplish, large corymbs.
LEMENHOF		Pink or blue, early.
LEOPOLD III	E. Draps, 1938	Deep pink or blue, late.
LE VENDÔMOIS	E. Mouillère et fils, 1932	Stout growth, rose pink. (Cert. Mérite Paris Exposition.)
LIBERTÉ	Lemoine, 1912	Pale pink, serrated sepals.
LILIE MOUILLÈRE	Mouillère, 1914	Deep pink, Solferino purple or deep blue, forces easily. A.M. 1914.
LORD LAMBOURNE	H. J. Jones, 1927	Deep pink. A.M. 1927.
LORELEY	Wintergalen, 1924	Crimson-pink or deep blue, early, short stout growth, free-flowering.
LOUIS FOUCARD	Foucard, 1912	Pink.
LOUIS MOUILLÈRE	Mouillère, 1920	Pure white, serrated sepals.
LOUIS PASTEUR		Deep pink, large corymbs, dwarf growth.
LOUIS SAUVAGE	Mouillère, 1928	Deep cherry pink or deep blue, dwarf growth, mid-season to late, inferior to 'Arthur Billard' for outdoors, but superior for pot-work, as it forces better.
LUCIFER		Clear red, early.
LUMINA	H. Cayeux, 1934	Pink, large flowers.
LUMINEUX	L. Mouillère, 1946	Massive stout wood, vigorous, bright Fuchsine pink. Late.
MLLE RENÉE GAILLARD	Chaubert, 1919	Milk-white, serrated sepals.
MLLE RENÉE JACQUET	Mouillère, 1920	Pink.
MME AIMÉE GYSELINCK		Red, early.
MME A. RIVERAIN*	Mouillère, 1909	Free-flowering, early, large, rather shapeless corymbs, pale pink or readily Cambridge blue.

Variety	Breeder	Remarks
		Often confused with 'Vibraye' but easily distinguished from it by the coloured petioles.
MME CHAUTARD (see 'SOUVENIR DE MME E. CHAUTARD')		
MME DE VRIES	D. Baardse, 1915	An *H. maritima* type, like a smaller edition of 'Joseph Banks', flowers greenish-yellow at first, then pale pink or pale blue in very acid soils.
MME E. MOUILLÈRE★★	E. Mouillère, 1909	White; pink or blue eye, tall growth, vigorous, serrated sepals, the best white Hortensia, but the corymbs turn pinkish with age, early and continuous flowering, bred from the Lacecap variety 'Whitewave' crossed with *rosea*. More long-lasting with light shade. The stems are pale green and not spotted. A.M. 1910. (See Plate F.)
MME FAUSTIN TRAVOUILLON★ ('PEACOCK')	Barillet, 1930	Cherry-pink or vivid mid-blue, vigorous and fairly free-flowering, early and late, does not spoil in sun unless wet.
MME FOUCARD	Foucard, 1912	Crimson-pink or deep blue, tall growth.
MME G. ALLERY	Mouillère, 1910	Bright pink or blue, Japanese type, free-flowering.
MME G. F. BIER	E. Draps, 1938	Deep pink or reluctantly mid-blue, mid-season flowering.

Variety	Breeder	Remarks
MME G. MORNAY	L. Mouillère, 1936	Stout strong wood, vigorous, large dark green leaves, warm pink serrated sepals.
MME GOUJON	Mouillère, 1912	White, large corymbs, late flowering.
MME HENRI CAYEUX	H. Cayeux, 1932	Crimson or port wine colour, stout growth of moderate height.
MME J. DE SMEDT	E. Draps, 1938	Stout, moderate growth. Pink or, if fed, pale Gentian blue (*H.C.C.* 42/2.) Mid-season, very free-flowering. Overcrowded bun-shaped heads. Requires part shade.
MME LEGOU	E. Mouillère, 1912	White, with a sheen of pink or blue, large heads, tall growth, not very free-flowering, late.
MME LUC CHAURÉ	Mouillère, 1923	Pink, serrated sepals.
MME MAURICE HAMARD	E. Mouillère, 1909	Pink, strong stems.
MME NICOLAUS LAMBERT	J. Lambert Söhne, Trier	Pink, large corymbs.
MME PAUL GIANOLI	L. Mouillère, 1938	Warm pink, serrated sepals, lasts well.
MME PHILIPPE DE VILMORIN	Mouillère, 1926	Deep pink, serrated sepals, free-flowering, moderate growth, late.
MME PLUME COQ		Pink, mid-season.
MME RAYMOND	E. Mouillère, 1909	Transparent white later going off pinkish.
MME RENAULT	E. Mouillère, 1910	Pink, rounded sepals, moderate growth.
MME TRUFFAUT	H. Cayeux	Pale pink or light blue, moderate growth.
MAGENTA	Lemoine, 1922	Pink.

Variety	Breeder	Remarks
MANDSCHURICA (NIGRA)		An old variety with black stems and pink or blue flowers. Mid-season.
MARCONI	K. Wezelenburg, 1936	Pink, serrated sepals.
MARÉCHAL FOCH★	Mouillère	Rich rose-pink, purple or vivid, deep, Gentian blue, very free-flowering, moderate to tall growth, readily gives a good blue but fades, early and continuous flowering; subject to mildew, has not a very strong constitution. A.M. 1923.
MARIE MATTHES	F. Matthes, 1924	Warm pink or pale blue, serrated sepals, moderate to strong growth.
MARS	K. Wezelenburg, 1936	Deep pink or purple, tall growth.
MATADOR	H. Cayeux, 1928	Crimson, stout growth, of moderate height, free-flowering.
MATILDA GUTGES		Blue, dwarf.
MAVIS		Crimson or deep blue, moderate growth.
MEIN IDEAL	F. Matthes	Bright pink or mauve, tall stout growth.
MEIN LIEBLING ('LIEBLING' OR 'DARLING')	F. Matthes	Warm pale pink, dwarf growth, early, does not blue readily.
MERVEILLE	H. Cayeux, Le Havre, 1927	Mid-season, vivid rosy-crimson, purple or mid-blue, large handsome corymbs, strong and massive but short growth. This variety is exceptional in having produced a number of branch-sports. Does not last well.

Variety	Breeder	Remarks
MERVEILLE BLANC	Dumas, 1937	White. Awarded medal, Paris, 1937. A branch-sport of Merveille.
MESDAG	K. Wezelenburg, 1927	Deep pink, moderate growth.
METEOR	Wintergalen, 1934	Crimson-pink or port wine colour, serrated sepals, moderate growth.
MIGNON	D. Baardse, 1915	Pink, dwarf growth.
MISS BELGIUM★	E. Draps, 1935	Rosy-crimson, mid-season, dwarf to moderate growth. Free-flowering.
MISS PHYLLIS CATO	H. J. Jones, 1927	Warm pale pink or blue, compact habit, early flowering.
MONSIEUR GHYS	Mouillère, 1912	Pink, large flowers.
MOUILLÈRE 723	Mouillère	White, less tall grower than Mme E. Mouillère.
MOUSSELINE★	Lemoine, 1909	Seedling of *rosea*. Pale pink or pale spectrum blue, flowers early and fairly continuously and remarkably freely, lasting well; tall, stout growth, particularly good outdoors with light shade or even in full sun. One of the best light blue outdoor Hortensias.
MRS. ALICE BLANDY	H. J. Jones, 1927	Crimson-pink.
MRS. A. SIMMONDS	H. J. Jones, 1927	Pink, serrated sepals. A.M. 1928.
MRS. BAARDSE (MEVROUW BAARDSE)	D. Baardse, 1918	Vivid warm pink (or, in very acid soil, blue), early flowering, stiff short growth, small plant. A.M. 1927. Not very free-flowering.

Variety	Breeder	Remarks
MRS. CHAS. DAVIS	H. J. Jones, 1927	Deep pink, serrated sepals, strong growth.
MRS. CHAS. MILLS	K. Wezelenburg	Warm pink, dwarf growth.
MRS. F. HUGGETT	H. J. Jones, 1927	Pink, large corymbs, vigorous.
MRS. H. J. JONES	Mouillère, 1923	Pale pink, serrated sepals, vigorous but dwarf growth. A.M. 1926.
MRS. L. J. ENDTZ	H. J. Jones, 1927	Pink, moderate growth.
MRS. R. F. FELTON	H. J. Jones, 1927	Large corymbs, pink or unwillingly deep purplish-blue; stout growth, moderate height. Mid-season. A.M. 1928.
MRS. W. J. HEPBURN	H. J. Jones, 1927	Large flowers, pink or 'deep purplish hue'. A.M. 1931.
MÜNSTER	J. Wintergalen, 1937	Mid-season flowering, velvety crimson, violet or deep blue flowers turning to strange and vivid tints in autumn, not very vigorous, dwarf growth, requires care and skill.
NEIGE ORLÉANAISE	Chaubert, 1922	A very fine white, holding its whiteness well without tinting; tall growth, rather tender. A.M. 1925
NIEDERSACHEN**	Wintergalen, 1914	Pale pink or, readily, blue, but needs shade to last well; tall growth; late.
NIKOLAUS LAMBERT	J. Lambert Söhne	Pink, strong growth.
NIXE	Wintergalen, 1933	Deep crimson or, if fed aluminium, intense deep blue, but does not last well; stout dwarf growth.
NORMANDIE	H. Cayeux, 1927	Cherry-pink, large corymbs.

Variety	Breeder	Remarks
ODIN	Wintergalen, 1932	Vivid rosy-crimson, strong, short, compact growth, large round corymbs, mid-season.
OPALE	Foucard, 1910	Blushed opalescent white, large corymbs.
ORIENTAL	Chaubert, 1922	Pink or, readily, mid-blue, large flowers but not very free-flowering until matured. Good near the sea.
ORIENTAL	H. Cayeux, 1935	('Rosebelle' × 'S. du Pres. Doumer'.) Strong stems. large cherry-pink flowers.
ORNEMENT	Lemoine, 1909	Purplish-pink, serrated sepals, large corymbs. A.M. 1910.
OSNING	Wintergalen, 1917	Crimson-pink.
OTAKSA	Imported from Japan	The Japanese Hortensia said to have been introduced from Japan by von Siebold in 1862,★ typified by orbicular, rounded leaves and short and free-flowering growth. Pink or blue flowers. Identical with Thunberg's original specimens which form the 'type' of the hybrid race of × *H. macrophylla*.
OTAKSA MONSTROSA	Lemoine, 1894	A branch-sport with larger flowers than 'Otaksa'.
PAPILLON	Mouillère, 1913	Yellowish-white with green.
PARADISIO		Salmon pink.

★Monsieur H. Decault, Secrétaire Général Fédération des Sociétés d'Horti-culture de France.

Variety	Breeder	Remarks
PARIS	Mouillère, 1928	Mid-season, deep pink or purple, tall and vigorous with very large flowers, free-flowering for one of the massive type.
PARURE	H. Cayeux, 1932	(Seedling of 'Merveille'.) Pink, rounded, crimped sepals.
PARZIVAL* (PARSIFAL)	J. Wintergalen, 1922	Rose Madder to crimson-pink or purple to deep blue, small numerous heads, serrated sepals, fades in sun quickly. Best if fed to give a deep Royal blue, but is variable in colour though, like other German varieties, the flowers last well in the shade and tint attractively in autumn. Moderate growth. A.M. 1922.
PASTEUR	J. Wintergalen	Deep pink, large corymbs. A.M. 1925.
PATRIOTE	H. Cayeux, 1935	Crimson-pink, compact heads, free-flowering.
PEACOCK (see MME F. TRAVOUILLON)		
PEARL OF SCHADEN-DORFF (see SCHA-DENDORFF'S PERLE)		
PEER GYNT	J. Wintergalen, 1921	Pale rosy-pink or lavender, serrated sepals, stout but moderate growth, mid-season.
PERFECTA	L. Cayeux, 1936	(Seedling of 'Merveille'.) Pink spherical heads, dwarf growth.
PERLE DU HAVRE	H. Cayeux, 1935	Massive stout growth, large serrated-sepalled pink flowers.

Variety	Breeder	Remarks
PETITE SŒUR THÉRÈSE DE L'ENFANT JÉSUS	Gaigne, 1947	Stout growth, white serrated sepals.
PHOEBUS		Pink, tall grower.
PIA		Dwarf pink.
PINK PRINCESS		Pink, early.
PORTA NIGRA	J. Lambert Söhne	Deep pink, strong growth.
PRÉSIDENT FALLIÈRES	E. Mouillère, 1910	Bright pink, vigorous.
PRÉSIDENT R. TOUCHARD	E. Draps, 1938	Cherry-pink. Very free.
PRÉSIDENT PINGUIET	Mouillère, 1909	Pink
PRÉSIDENT POINCARÉ	Mouillère, 1914	Pink.
PRÉSIDENT VIGER	Mouillère, 1909	('Vibraye' × 'Otaksa'.) Pink, vigorous constitution.
PREZIOSA★★		Hybrid, flowers turn red in full sun, slender growth. Good outdoors, see p. 137.
PRIMA	E. Draps, 1938	Deep pink, mid-season.
PRINCE BLUE (see 'BLAUER PRINZ')		
PRINCE HENRY		Sport of 'Parzifal', crimson-pink, dwarf growth.
PRINCE ROUGE		Red, early.
PRINCESS BEATRIX★	E. Draps, 1946	Crimson-pink to purple; moderate growth, early to mid-season, very free-flowering. Flowers turn red in autumn.
PRINCESS ELIZABETH	H. J. Jones, 1929	Pink or blue, tall growth.
PRINCESS JULIANA★	D. Baardse, 1921	Creamy-white, later blushing pink, mid-season to late-flowering, entire rounded sepals, dwarf growth.
PRINCESS MARINA	H. J. Jones, 1927	Pink.

Variety	Breeder	Remarks
PROFESSEUR D. BOIS	E. Mouillère, 1909	Warm pink, large corymbs, rounded flowers, late. A.M. 1922.
PROFESSEUR DE VRIES		Pink, moderate growth.
PROFESSEUR LENFANT	1938	Large pink corymbs.
PROGRÈS ·	Lemoine, 1915	Cherry-pink or deep blue; flowering over a long period.
QUEEN EMMA	D. Baardse, 1920	Crimson, mid-season, tall growth.
QUEEN MARY	K. Wezelenburg, 1927	Pink. A.M. 1927.
QUEEN WILHELMINA	Lemoine, 1910	Seedling of *rosea*, pink, readily blue, mid-season.
RADIANT	Lemoine, 1910	Rose-pink or readily pure blue, bushy habit, mid-season. A.M. 1915.
RAYMOND DRAPS		Pink, mid-season.
RED EMPEROR	E. Draps, 1938	Vivid rosy-crimson, late flowering.
RED STAR		Red, mid-season.
REGULA	Moll, Zürich, 1934	White, strong stout growth, mid-season.
REINE DE BEAUTÉ	L. Cayeux, 1936	Pink, large compact corymbs.
RÉNOVATION	H. Cayeux, 1930	Early to mid-season, rich crimson-pink, tall growth.
REX	H. Cayeux, 1935	Seedling of Président Doumer, crimson, large flowers.
RHEINGOLD	J. Wintergalen, 1920	Bright pink, compact moderate growth.
RHEINLAND		Red, mid-season.
RICHESSE	Lemoine, 1912	Vivid pink.
ROCHAMBEAU	Mouillère, 1918	Crimson or violet, or deep blue if fed, rounded sepals, moderate to dwarf growth, mid-season to late.

Variety	Breeder	Remarks
ROI ALBERT	Mouillère, 1914	Pale bluish or pinkish-white, large flowers with serrated sepals.
RONSARD	Mouillère, 1909	Similar to Mme Mouillère but blushed with pink or bluish tint.
ROSA KING		Pink, mid-season.
ROSALINDE	Lemoine, 1920	Pink, serrated sepals.
ROSEA★	Imported from Japan	A Japanese hybrid Hortensia clone distinct from Otaksa having domed, more elliptic leaves and a more dome-shaped corymb. One of the four principal ancestors of the European bred Hortensias. 'Rosea' was introduced to England by Charles Maries for Messrs. Veitch in 1880. Loose heads, pale pink or blue. It takes time and some initial feeding to give good blue flowers. 'Rosea' is unusually hardy and free-flowering out-of-doors with only moderately tall but vigorous, slender growth.
ROSEBELLE★	H. Cayeux, 1923	A good pink, mauve or blue, moderate growth, rather erect grower, hardy, good in Australia, mid-season.
ROSE DE TOURS	Barillet	Pink, spherical corymbs.
ROSITA		Pink, strong grower.
ROUGET DE LISLE	Lemoine, 1926	Crimson, purple or deep blue, compact rather overcrowded heads,

Hydrangea macrophylla 'Hamburg'

Hydrangea macrophylla 'Vulcain'

Hydrangea macrophylla 'Westfalen'

X hydrangea macrophylla 'Mme E. Mouillère'

Variety	Breeder	Remarks
		dwarf growth, very free-flowering, large wavy sepals. An excellent variety in France, little known here under its correct name.
RUBENS	K. Wezelenburg, 1936	Deep pink.
RUBIS	Lemoine, 1923	Ruby-red, rosy flushed maroon or purple, moderate, rather weak, compact growth, requires care.
RUTILANT	L. Cayeux, 1936	('Flambard' × 'Vésuve'.) Crimson flowers on strong stems on a bushy plant.
SACHSENKIND	F. Matthes	Pink, moderate growth.
ST. BONIFAZ	F. Matthes	White, serrated sepals, early, strong growth.
ST. CLAIRE		Blue or pink, mid-season.
SAMBA		Red, mid-season.
SANGUINEA	Albert Truffaut, 1938	A branch-sport of 'Merveille' with blood-red flowers.
SATINETTE	Lemoine, 1916	Crimson-pink, tall growth.
SCHADENDORFF'S PERLE	H. Schadendorff, 1928	Pink or blue. Does not last when cut. Tall growth, large sepals, early.
SCHÖNE DRESD-NERIN (BEAUTY OF DRESDEN)	F. Matthes	Vinous pink, moderate growth.
SEDGWICK'S WHITE (syn. MME E. MOUILLÈRE)		This is apparently only an English name for 'Mme E. Mouillère'.
SEESTADT WISMAR	Paul Kiske	Warm pink, strong growth.
SÉNATEUR HENRI DAVID	Lemoine, 1910	Pale pink.

Variety	Breeder	Remarks
SENSATION		Pink, tall growth.
SIBILLA		Deep rosy-red, late.
SIEGER	J. Wintergalen	Early, pink, tall growth, serrated sepals.
SIEGFRIED	J. Wintergalen, 1926	Deep pink, moderate growth.
SIGYN HARTMAN	J. P. Hartmann, Ghent, 1935	Clear pink or clear blue, medium height. Good in Australia. Moderate growth.
SIROCO		Deep pink, mid-season.
SOEUR THÉRÈSE		White, poor outdoors.
SONNENEGRUSS	F. Matthes, 1936	Soft cherry-pink or bluish-purple, very compact heads, moderate growth.
SOUVENIR DE CLAIRE	Mouillère, 1908	Pink, offspring of *rosea* and very free-flowering as it flowers on the side-shoots also, but the foliage is weak and subject to mildew under glass.
SOUVENIR DE MME BÉRANGER	Barillet, 1913	Flowers pink, paler centre, bushy habit, dwarf growth.
SOUVENIR DE MME E. CHAUTARD	Mouillère, 1909	Clear pale pink, mauve or blue, moderate growth. An unreliable variety in colour, often throwing both pink and blue flowers at the same time; early flowering. Bred from 'Souvenir de Claire' × *rosea*.
SOUVENIR DU PRÉSIDENT DOUMER★	H. Cayeux, 1932	Shapely flowers, intense dark velvety red or purple or dark blue, dwarf growth. An indispensable variety for the favourable

Variety	Breeder	Remarks
		garden but requires care. Mid-season.
SPÄTSOMMER	F. Matthes	Pink, short, stout growth, late.
SPLENDENS	H. Cayeux, 1920	Clear pink or clear blue readily obtained; stiff growth of moderate height.
STARLET		Deep pink, mid-season.
SUCCÈS	H. Cayeux, 1920	Pink, serrated sepals.
SUPERBA	H. Cayeux, 1934	Large pink flowers.
SUPERSTAR		Red, early.
SURPRISE	Lemoine, 1911	Greenish-white.
SUZANNE CAYEUX	H. Cayeux, 1919	Pink (Certificate of Merit, Paris Ex.).
SWANHILD	Wintergalen	White, flushed pink, moderate growth, early.
TERRE DE FEU	Lemoine, 1919	Cherry pink.
THOMAS HOGG	Imported from Japan	One of the first importations from Japan, this Hortensia shows hybrid origin. The shoots are very thin and weak and the head is packed with pure white sterile flowers of irregular shape, most having long, pointed, entire sepals. A branch-sport of 'Lanarth White'. (See Illus. 21.)
THOMAS STEVENSON	H. J. Jones, 1927	Pink, serrated sepals.
TÖDDY		Red, mid-season.
TOSKA (TOSCA)	Wintergalen, 1930	Pink or pale blue, doubled flowers, serrated sepals, vigorous growth, moderate height, stout bushy habit. Mid-season.
TOURAINE	Belenfant, 1938	Sport of 'Merveille', pale pink or pale blue, sepals waved, lasts well.

21 'Thomas Hogg', the first white Hortensia, imported from Japan in the nineteenth century

22 *Hydrangea macrophylla* var. 'Mariesii', introduced in 1879

Variety	Breeder	Remarks
TRIOMPHE	H. Cayeux, 1920	Vivid pink or mid-blue; free-flowering, compact growth, mid-season. A.M. 1923.
TROPHÉE	Lemoine, 1915	Rich deep pink, large flowers, dwarf growth.
UNIVERSAL		Red, mid-season.
URSULA		Early, blue.
VENUS	K. Wezelenburg, 1936	Clear pink, mauve or blue, tall growth.
VÉSUVE	H. Cayeux, 1932	('Champion' × 'Merveille'.) Deep rich pink or mauve to deep blue, serrated sepals, stout massive growth.
VIBRAYE (see 'GÉN. VIC. DE VIBRAYE')		
VICTOIRE	H. Cayeux, 1920	Apple blossom pink.
VIEUX CHÂTEAU	Mouillère, 1909	Pink.
VIKING (see 'WIKING')		
VILLE DE VENDÔME	E. Mouillère, 1910	Creamy-white, coloured eye, large corymbs, vigorous.
VIOLETTA★★	Haworth-Booth	Cherry red or violet, long-lasting flowers, hardy.
VULCAIN★★	Henry Cayeux	Full-sized corymbs of shapely flowers, crimson, purple, or orange and green on a dwarf plant a foot high. Not less hardy than others but needs time to grow; worth the attention of the skilful grower owing to the fascinating colours obtainable in soils of varying acidity. Continuous flowering all season. Good in sun, but

Variety	*Breeder*	*Remarks*
VULCAIN** – *cont.*		sometimes reverts to tall growth. (Plate E.)
W. A. MOWBRAY	H. J. Jones, 1927	Pink.
W. D. CARTWRIGHT	H. J. Jones, 1927	Bright pink, dwarf habit.
W. J. HEPBURN	H. J. Jones, 1921	Pink.
WESTFALEN**	Wintergalen, 1940	Mid-season, vivid crimson or violet, beautifully shaped flowers, dwarf to moderate growth. Flowers often on the shoots of the year in the same season if well cared for. Award of Merit, 1958. (Plate G *f.p.* 115.)
WESTFALENKIND	Wintergalen, 1913	Pink or blue, early. Does not fade in sun, poor growth in shade.
WIESBADEN	Moll, Zürich	Deep pink, purple or blue, dwarf growth.
WIKING	Wintergalen, 1932	Early, pink, serrated sepals, moderate but compact growth.
WILLIAM PFITZER	Anton Rosen-kränzer, 1913	Pink or blue, moderate growth.
WRYNECK		Downy leaves with red stalks and veins. The crowded corymb of sterile flowers is pink or deep cobalt blue. Unfortunately the stem twists in a curious manner so that the flower-head does not long hold itself up properly. Imported from Japan. Similar to 'Otaksa'.
YVONNE CAYEUX	H. Cayeux	Deep pink, tall.
ZUKUNST	F. Matthes, 1928	Pink, moderate growth.
ZÜRICH	Gebr. Moll, 1947	Strong variety, clear rose,

Variety	Breeder	Remarks
		dark green foliage, strong stems, substitute for 'La Marne' for pot-work.

One may well wonder why these numerous new Hortensias are produced every year by simply sowing the seed of the old varieties. Where should we be with Rhododendrons if the breeders had merely sown the seeds of 'Doncaster' and 'Pink Pearl' every year instead of hybridising with innumerable different wild species to produce quite new qualities?

The new Hortensias produced are, as one would expect, all very much the same as the older sorts, with so little individuality indeed that I do not know of any nursery where all the varieties are correctly labelled. The answer to the question is, I am told, that the Hortensia buyers *insist on new names*. In reality it would be much better quietly to rename the best of the old sorts – 'Marshal Foch' could come out as 'Blue Dream'; 'Ami Pasquier' as 'Scarlet Passion'; 'Kluis Superba' as 'Mood Indigo', and so on. Indeed, this has sometimes been done already as, for example, when 'Heinrich Seidel' was re-issued as 'Glory of Aalsmeer'. But the breeders are honest fellows and good businessmen for the most part, and they are simply meeting the demand. For many years I have tested all the new sorts but they are simply not free-flowering enough outdoors and the flower quality shows no decisive improvement. This does not matter, for they are not produced for use as outdoor flowering shrubs. On the other hand, the old double-starred varieties are splendid garden decorators. Indeed, they are more effective flowering shrubs than even the finest Rhododendrons. Instead of a flower-life of eight days we have – in a variety like the superb 'Altona' – four or five *months* of flower beauty besides fine autumn leaf colouring. Blue is one of the most greatly enjoyed colours and 'Vibraye' provides the most glorious masses of pure soft 'Butterfly Blue' in myriads of gardens far better than any other plant can do. Deepest of all the blues I have grown is 'Gentian Dome' (Enziandom★) and

★ Translation of foreign names is permitted by the Rules.

for outdoors it supersedes 'Marshal Foch'. Where the 5 foot 'Vibraye' is too big, the best substitute is 'Fisher's Silverblue' – 2 feet 6 inches. Of the pinks 'Hamburg' is my choice, with 'Violetta' as a deeper cherry pink. Of the red-flowered Hortensias none quite approaches the hardiness and freedom of flower of 'Westfalen'. 'Europa' is a readier bluer than 'Altona'; indeed, this fact and the more open and less massive flower-head are the chief distinctions during the summer, but in autumn the blazing pure red to which pink, lilac, or blue 'Altona' flowers turn, makes it the better variety. White Hortensias are led by the superb 'Mme Emile Mouillère'; 6 feet high and successionally and perpetually flowering, it is unbeatable as a north wall decorator. But in the garden beds a sun-exposed and carefully starved 'Lanarth' Lacecap gives a better return.

As for the dwarfs – alas, the lovely 'Vulcain', evidently a branch-sport, often reverts disastrously to a tall, shy-flowering variety. 'Pia' has a poor dull pink flower, and 'Hornly' seldom flowers at all.

For Hydrangeas grown as outdoor flowering shrubs, freedom of flower is the first essential. The double starred varieties are so free-flowering that every branch usually carries several corymbs and new shoots will often flower in the same season, like a herbaceous plant. Thus, if the top is killed by frost, flowers may still be had, provided that any excess numbers of new shoots are removed and frost-damaged wood is carefully cut back to healthy growth. Regular feeding is also, of course, essential to promote the rapid growth required.

The enormous number of pink-flowered varieties, is so bewildering as to make selection very difficult. The commercial grower of plants for indoor use has quite different requirements to those of the gardener requiring outdoor flowering shrubs. From this latter angle matters can be easily simplified. There are really only four colours available for either type of soil (acid, or neutral to alkaline). In acid soil we may, however, nearly double the number by neutralising the surrounding soil for certain plants which we wish to prevent showing blue flowers. Thus in acid soil (pH 4 to pH 6) we may have white, pale pink, deep pink, purple, crimson, pale blue, deep blue or deep purple flower colours by mere minor attentions to feeding the individual plants. In a limy soil we have only white, pale pink, deep pink and crimson flower colours available, unless we are prepared to take considerable trouble in continually feeding the plants with the required mineral food which they cannot otherwise obtain from such a soil.

Thus from the colour point of view, if it were necessary to limit the number of varieties to the absolute minimum, eight varieties would practically cover the available range. Admittedly there are some pinks optimistically described as 'salmon' which are warmer in tone than others. They do not, however, match the 'Azalea Pink' of the *Horticultural Colour Chart* as real 'salmon' should, being really cherry or crimson pink which is often called peach-pink when pale. This is a more pleasing colour to most eyes than Rose Madder pink, which is the 'cake-icing' pink of many of the old varieties when grown in neutral or alkaline soils. Most of us will wish to avoid such hues and select the warmer 'flesh' and cherry pinks. Thus matters are already simplified because we have only to select the most free-flowering, hardy, vigorous and healthful varieties of these colours.

Depending on the relative strength of the *H. japonica* non-blueing strain in their composition, the readiness of the varieties to attain a reliable pure blue colour varies widely. Those whose colour genes derive from *H. acuminata* and, to a lesser degree, from *H. maritima* naturally settle down quickly in a properly acid soil. Others, having the *H. japonica* colour genes present, lack the valuable quality of ready assimilation of aluminium and give flowers of uncertain purplish hues greatly inferior to the beautiful pure blue colouring of the best sorts, which is unrivalled in purity by any other flower of comparable size that can be grown out-of-doors in the British climate. Thus, in selecting our blue varieties, these points should be borne in mind.

If we have sufficient space we may wish to cater for another angle – the succession of bloom. In this case we shall want an early (late June or early July) flowerer, a mid-season (late July) flowerer and a late-flowering (August) variety in each colour. Thus the various colours may each be represented by three varieties. This would amount to about a couple of dozen sorts. In practice, a number of other varieties notable for having unusual flower shapes – serrated sepals or a corymb of unusual form – will be desirable as well.

I think we may take it, therefore, that allowing for the finer nuances of hue as well, about fifty varieties out of the three hundred or more listed would comprise sufficient worthwhile sorts of outdoor growing. A number of varieties of proved excellence in my experience and which are available in commerce are starred, two stars being given to outstanding varieties most suitable to form the main mass of the plantings. These will, I think, offer the gardener a sound foundation selection for a start. Novelties and other kinds can then be easily judged as to their relative merits in comparison with these.

THE LACECAPS

In this section the flower-head, instead of being globose and almost wholly formed of sterile flowers, is a comparatively flat, roughly disc-shaped corymb. The central area of small fertile flowers is surrounded by a marginal ring of large, showy, sterile ray-flowers.

This is one of the most unusual and beautiful shapes for an inflorescence. Some members of the Viburnum family, including our native Guelder Rose (*V. opulus*), share this charming arrangement and, of course, most of the wild species of Hydrangea.

Compared with the vast numbers of varieties in the Hortensia section there are only a very few of these hybrid Lacecap varieties but they are so little known and so valuable as outdoor flowering shrubs that more detailed description is deserved.

In woodland gardens reasonably free from slugs, where they may succeed the Azaleas, Rhododendrons and Rose species in bloom, the Lacecaps are superb. Their remarkably strong and vigorous growth enables them to make large permanent bushes, untended for years, where the weaker Hortensias would soon fall a prey to pests and enveloping weeds. As less attention has been given to their breeding, and because the *H. japonica* strain is more dominant and the vivid *H. Thunbergii* strain almost absent, they lack the intensity of colouring of the choicer Hortensias, offering us only relatively pale tints, but such is the admirable form of the flower-heads that beautiful lace-like effects are produced.

Owing to the name muddles regarding × *H. macrophylla* var. 'Mariesii' and the offspring of this garden variety, already described in our chapter on the History of the Garden Hydrangea, I have had to propose names for some varieties which had no valid name. With the Hortensia section there were often more names than plants. With the flat-headed Lacecaps it was the opposite. When I first tried to sort them out many years ago there were not enough names for the number of quite distinct clonal varieties found growing in gardens all over the country. A variety that I have not seen and described below may possibly come to light, but I have hunted them assiduously for so long that I think that all are now included.

The following list is arranged as nearly as possible in flowering date order:

×**Hydrangea macrophylla 'Lanarth White'** (A.M. 1949). This lovely variety, of which huge specimens still grow in Mr. Michael Williams's superb garden at Lanarth, Cornwall, is a compact but rather slender-branched bush of 2 or 3 feet as a rule. The leaves are a more yellowish-green than those of most Hydrangeas unless iron is artificially fed and the stems are unspotted pale green. The flat flower-head is composed of a central circular area of bright blue (or pink in limy or neutral soil) fertile flowers with a marginal ring of absolutely pure white ray-flowers. The sepals are pointed and entire and usually four in number. Thus the effect is that of a diadem of

23 *Hydrangea macrophylla* var. 'Tricolor'

blue and white (or pink and white) delightfully framed by the vivid foliage. 'Lanarth White' revels in full exposure to sun and wind in the most unusual way and will make do with very poor soil. The most attractively grown specimen that I ever saw was in a very poor soil where it grew right in the open as a dense mound 3 feet high so covered with flowers that hardly a leaf could be seen. It is thus an ideal shrub for the small garden. It flowers with the first Hydrangeas to open, being often at its best in early July. So long-lasting are the flowers, however, that it remains attractive until the latter part of August. In shaded places with a rich soil it is much less effective as the ray-flowers become irregular in shape. The best Hydrangeas for such positions are some of the varieties of the *H. serrata* group such as will be described later in their own section. Overfed, 'Lanarth' sports into 'Thomas Hogg'.

'Lanarth White' is one of the rarest of the Lacecaps though one of the most valuable. I have been unable to trace the origin of this Hydrangea with certainty from any published description or figure. It is possible that it might be the 'Shirogaku' mentioned in our historical chapter. Again, it has some resemblance to the variety *cœrulea* figured in the *Bot. Mag.* (Hooker, t. 4253, Vol. 72 [1846]), but this is described as having the ray-flowers blue at the base, and the ray-flowers of 'Lanarth White' are absolutely pure white. It is the whitest Hydrangea flower that I know, and an unusually hardy variety of puzzling origin.

×**H. m. 'Lilacina'** (*H. Mariesi lilacina* Lemoine). This variety was bred by Messrs. V. Lemoine of Nancy and brought out in 1904. It was a chance-pollinated seedling of *H. m.* var. 'Mariesii'. It has a distinct *H. japonica* hybrid appearance, clearly indicating the hybrid origin of its parent. Lilacina is a rather coarse, strong-growing plant with large pointed leaves and a comparatively small flower-head of the usual Lacecap shape with the four-sepalled ray-flowers boldly serrated at their edges. In colour it varies (exactly, of course, in accordance with the acidity, or pH value, of the soil) from a clear phlox pink through lilac to pure blue. With age, in acid soil and shade, *lilacina* loses the disproportionately large foliage and flowers more freely, becoming a very handsome Hydrangea. In a neutral soil it is not so attractive in the early stages as others. It does best in light shade.

×**H. m. 'Veitchii'** (*H. Mariesii* var. *Veitchii*). This variety is fairly common and has very large ray-flowers with three entire, white sepals. Exhibited by Messrs. Veitch of Exeter at the Temple Show in 1903 under this varietal name and described in their *Catalogue of Hardy Trees and Shrubs* in 1905, p. 42, it was soon widely distributed and was used by the French breeders in the production of the Hortensias. 'Veitchii' is obviously a hybrid, having, indeed, quite as much of the woodland species in its composition as the maritime. It seems necessary to disregard Wilson's taking of the name 'Veitchii' in 1923 to foist it upon the well-known, fully authenticated foundation clone 'Rosea',★ apparently in order to take the name *rosea* for another Hydrangea. 'Veitchii', then, forms a tall bush up to 6 feet in height in shade and rich soil. It is rather an untidy grower having very rich green, notably acuminate leaves, often puckered and bullate and small flat corymbs with unusually large white ray-flowers which age to a pinkish tint in sunlight and a greenish one if in the shade. The flowering is about mid-season. In a fully exposed sunny position, 'Veitchii' grows poorly and is inferior to 'Whitewave' or 'Lanarth White' and in a very densely shaded position it is perhaps slightly less effective than *H. japonica macrosepala*. On the other hand, in light, dappled shade it is a very beautiful Hydrangea, although a martyr to slug attacks. (See Illus. 24.)

×**H. m. var. 'Mariesii'**, A.G.M.† This beautiful hybrid garden variety from Japan is securely documented and identified beyond any doubt. Yet it is one of the most seldom seen,

★ *Revue Horticole*, Paris, p. 544 (1904), and *Gartenflora*, t. 1533 (1904), and *Hortus Veitchii*, 368 (1906).

† The Award of Garden Merit. This important distinction is only awarded to a plant after trial at the Royal Horticultural Society's gardens to make certain that the plant is a good garden plant likely to give satisfaction as such in members' gardens.

or correctly named. Introduced from Japan by Charles Maries for Messrs. Veitch of Exeter in 1879, there are still fine specimens (from cuttings several times no doubt) of the original plants at Kew. The inflorescence, opening at mid-season, is bun-shaped, almost midway in form between that of the Hortensias and the Lacecaps. Most of the small fertile flowers are concentrated in the centre of the corymb, but, in addition to the large showy ray-flowers around the margin, abortive smaller ones of this shape are scattered about the centre also. This gives the bush an attractive informal appearance when in flower. The sepals of the ray-flowers are usually four in number and rounded and entire in outline. In colour the flowers are a pale rose-pink; only after some years' growth in very acid conditions or when aluminium is freely fed to the plant do they finally come a pale blue. Indeed, I have yet to see a really pure blue flower-head of this variety. 'Mariesii' is only of moderate height – usually about 4 feet. It is by no means common and I only saw a very few plants during two tours of the west country, hunting Hydrangeas. 'Mariesii' has unusually narrow and evenly tapered leaves and thus can be easily distinguished when out of flower. (See Illus. 22.)

×**H. m. var. 'Whitewave'** (A.M. 1948). This outstandingly beautiful variety is also so well documented that there is no doubt as to its identity. Like 'Lilacina' and 'Bluewave' it was raised by Messrs. V. Lemoine of Nancy about 1902 from naturally pollinated seed taken from a flower-head of *H. m.* var. 'Mariesii'. It was brought out in 1904 under the name of *H. Mariesi grandiflora*. This name not being in accordance with the rules of taxonomy I ventured to propose for it the simpler and more descriptive name of *H. m.* var. 'Whitewave' duly published as cited below.

'Whitewave' received the Award of Merit when exhibited in 1948 at the R.H.S. (subject to verification of name which was duly confirmed later) and was well figured in *The Gardeners' Chronicle* (3217, **124:** 77 [1948]), and *Gardening Illustrated* (October, 1948, p. 196) and the *Journal, R.H.S.* (LXXIV, **5:** t. 59 [1949]). The growth is strong and vigorous, the shrub reaching in time to about 5 feet in height and width. The leaves, noticeably spoon-shaped at first, are thick and elliptic with an acute tip. The flat flower-head is symmetrical and a well-defined central disc of blue (or pink) fertile flowers and about eight very large and beautifully formed pearly, faintly blushed, white ray-flowers regularly spaced around the margin. The four sepals of each of these are delicately waved, or toothed, at the margins and the corymbs are presented in such an extraordinarily free and regular manner that a bush in full flower is one of the most decorative sights in the garden in the middle of the season. In an open position every side shoot carries a flower-head. In shade it is less free-flowering and compact in habit. (See Illus. 25.)

×**H. m. var. 'Azisai'** ('Yodogama'). This old Japanese garden variety is one of the best known in that country, being much prized, according to von Siebold, for planting in gardens and temple groves, but I only discovered later that it had been imported here long since and, at all events, is apparently present in certain British gardens today. Generally known as 'Yodogama' in this country, it was listed in the superb pre-emergency catalogue of Messrs. V. N. Gauntlett of Chiddingfold. It was originally imported, as legend has it, by a sea-captain who, wanting a pot-plant for his wife, at some time during the closure of Japan to foreigners (1638–1856), sent a native hand ashore at Yokohama to get one. The lad returned with this Hydrangea which ultimately duly reached England.

Superficially this clone (or vegetatively propagated garden variety) resembles 'Bluewave', described later, but the leaves and the flower-head are differently shaped. The leaves have the more orbicular, rounded outline of the Japanese clone 'Otaksa' and the flat flower-head is markedly unsymmetrical and irregular. The ray-flowers around the margin have three or four variously-shaped sepals, seldom waved at the edges. In colour they are pink, purplish or a pale tint of Gentian blue. A characteristic of this variety is the unusual length of the pedicels (or flower-stalks) of the ray-flowers, these being over 2 inches long.

24 *Hydrangea macrophylla* var. 'Veitchii'. An important hybrid foundation clone imported from Japan in 1902

25 *Hydrangea macrophylla* var. 'Whitewave'

I have not yet had time to test this variety out fully as an outdoor shrub. At first sight it seems inferior to 'Bluewave' in every way, but it is possible that the foliage and flower colour may resist the action of the sun's rays better. If this is so, it will be found to be a useful addition to the Lacecap range for sunny positions.

×**H. m. 'Belzonii'.** E. H. Wilson cast doubt upon the existence of this variety, but it does exist over here in exact accordance with Siebold's figure★ and has such a clear-cut botanical distinction in the fact that the leaves are produced in threes instead of pairs on the shoots that we cannot ignore it.

Belzonii is not a very decorative variety unless grown under very acid conditions, or fed aluminium, and positioned so as to receive light shade. It appears to be a hybrid of very mixed parentage. The leaves are not invariably in threes but are usually so, and the flat flower-head has about five or six four-sepalled pink, lilac or blue ray-flowers of no great size or beauty. It is of more interest to the collector than the gardener. In the past some writers have mentioned the 'remarkable beauty' of its flowers, but I think that, in such cases, they have confused its identity with 'Azisai', 'Bluewave' or even 'Bluebird'.

×**H. m. 'Quadricolor'.** I found this fine variety in the renowned, superb garden of General Harrison at Tremeer. The leaves carry four distinct colours – pale green, cream, deep green and vivid yellow. The flower-head is quite attractive and of Lacecap type, pale pink. It makes quite a strong bush 5 feet high in shade. Slugs are rather overfond of the young growth so care has to be taken to put cinders round the stems and slug poison close by or growth will be crippled. It is a favourite with flower arrangers who do wonders with its lovely leaves.

×**H. m. 'Tricolor'.** This is a very fine variegated variety, the leaves showing three colours, deep green, sea-green and pale yellow. The flower-head resembles that of *H. m.* var.

★ *Flora Japonica,* **1:** 109, t. 55 [1840].

'Mariesii', but is perhaps slightly paler in colour. Some gardeners object to variegated foliage but those who do not share this view will find *tricolor* one of the most attractive of all shrubs of this kind. In addition, it is remarkably free-flowering over a long period and very vigorous in growth, making a bush 6 feet high under good conditions. If fed aluminium the flowers are a very decorative pale blue colour. As a tub or cold-house plant I think this Hydrangea has been undeservedly neglected. It was evolved in Italy by Messrs. Rovere of Pallanza being, it is thought, a branch-sport of 'Mariesii'. It was first shown at the R.H.S. in London by Mr. Bull, a nurseryman of Chelsea, in 1862. (See Illus. 23.)

×**H. m. var. 'Bluewave'.** This variety was bred by Messrs. V. Lemoine of Nancy from seed of × *H. m.* var. 'Mariesii', naturally pollinated. It was brought out in 1904 under the name of *H. Mariesi perfecta* and was widely distributed in England. Being an unusually strong and hardy variety it formed a large permanent bush wherever it was planted. The labels did not last as long as the plant and soon, naturally enough, all of the name that the gardeners could remember was *Mariesii*. Thus a distinct seedling variety became known only by its mother's name. It is precisely in order to prevent such an improper state of affairs that the rules of nomenclature most rightly forbid the carrying, by a distinct variety of a given species, of a name already held by another.

At the time, Messrs. Lemoine no doubt considered 'Mariesii' to be a species, not merely a garden variety, and in those days the naming of plants was not so carefully considered as it is today. In view of the muddle caused by a name not permissible according to the rules I ventured to rename this variety × *H. m.* var. 'Bluewave'. 'Bluewave', then, is one of the finest of all Hydrangeas as an open-air garden shrub in light shade. It forms a strong rounded bush 6 feet or more high and through.* The branches are stout and the leaves elliptic with an acute point and the flower-head is symmetrical and

---

* I recently measured a bush, solidly laced with blue flowers, 24 feet in diameter and 8 feet high.

beautifully shaped. The circular central area is filled with fertile flowers and the numerous, regular and large ray-flowers are four-sepalled with attractively waved edges. In colour they are pink, lilac or a pale tint of Gentian blue.

'Bluewave' requires markedly acid conditions or careful and gradual aluminium feeding to show its best colouring – it is not a remarkably ready bluer, like 'Vibraye' or 'Mousseline' and is easily upset and severely damaged if given too strong and sudden a dose. Furthermore, in too sunny a position the young leaves scorch and the flower colour easily bleaches to a muddy tint. On the other hand, when well grown in light shade there are few flowering shrubs of any kind which can equal the beauty of a fine specimen of this Hydrangea. It is the commonest of the Lacecap varieties, immense plants being found in old gardens in all parts of the country. So long as the bushes are not interfered with by man they appear to be quite permanent, but many are killed by being cut down after flowering. Soft shoots galore then spring up, and, if an untimely autumn frost destroys them, the plant is crippled for years. During this time weeds may choke it and slugs destroy every basal shoot and thus it dwindles and finally dies out, a victim of thoughtless 'pruning'. 'Bluewave' has the characteristics of a hybrid containing 'blood' of all the basic types, *H. maritima* being perhaps the most favoured. (See Illus. 18.) Award of Merit 1956. First Class Certificate 1965.

×**H. m. var. 'Beauté Vendômoise'.** This beautiful variety
was bred by Messrs. E. Mouillère et Fils in 1908, and is unfortunately now very scarce. The individual sterile flowers are the largest attained, I believe, by any Hydrangea, being described by Monsieur Louis Mouillère, in a letter to me, as often attaining 12 centimetres (about 4¾ inches) in diameter. The fact that, as with the variety 'Mariesii', many fertile flowers also filled the corymb detracted from its value for the florist trade *of those days*. My italics are to make it clear that our florists of today are not so misguided; indeed, I have been much surprised at the keen interest they have shown whenever I have exhibited the Lacecap varieties.

Monsieur Mouillère has kindly sent me some cuttings from the surviving stock plant at Vendôme so it has been possible to test it under outdoor conditions here. It is still grown and prized in Australia. In colour the flowers are white, blushed with carmine, and the growth is moderate.

The foregoing clonal garden varieties comprise the Lacecaps coming under the head of the hybrid race of *H. macrophylla* which contains a proportion of the 'blood' of *H. maritima*. Ere long we shall see a number of new seedling varieties and it will be necessary to judge their merits strictly in comparison with those which we already have. As will have been realised from the descriptions given, there is a dearth of hardy Lacecap varieties with vivid deep-coloured flowers able to grow in full sun and exposure. A variety with the splendid constitution and free-flowering nature of 'Whitewave' coupled with the rich flower colouring of 'Président Doumer' or 'Westfalen' would be a most desirable acquisition.

To assist in the correct naming of the varieties I am endeavouring to form as complete a collection as possible true to name. Situated at Farall, Roundhurst, near Haslemere, it is now open to inspection by interested persons.

Chapter 4

Hydrangea Japonica and acuminata varieties

I feel that I must retain Siebold's name *japonica* for the wild species, which I have more recently obtained from Japan, its varieties having the curious characteristic of white flowers turning vivid red under the action of sunlight. This is coupled with a rather weak constitution and the necessity for shaded growing conditions with moist soil. They are notably much more difficult and fastidious than the *acuminata* varieties.

I supplied the National Herbarium and the Wisley Herbarium with a set of specimens.

H. japonica 'Macrosepala'. This is one of the most distinct of the forms of the various parental species of the hybrid race. It definitely requires a cool, shady situation and moist soil. A slightly built bush of 70 to 90 centimetres it bears flower-heads carrying usually only four or five large deckle-edged ray-flowers of a pure white that turns to vivid crimson if exposed to sunlight. I shall never forget my excitement when I first set eyes on this beautiful plant under a purple plum bush in the gardens of Trewidden, Penzance. I was kindly given cuttings and have grown it ever since. See Illus. 19. It was unmistakably illustrated in *Gartenflora* **15**: 290, t. 520 [1860].

H. j. 'Impératrice Eugénie'. This variety (Carrière, *Revue Horticole*, 469, t. [1868]) has a small Hortensia-type flower-head of white sterile flowers that turn vivid red by the action of sunlight. Like many Hydrangeas whose flower-stalks were made to hold up the lighter lacecap-type flower-heads these heavier corymbs are apt to droop when weighted by rain. Until I discovered its true name I grew this plant under the 'kennel name' of 'Hohenlinden' in allusion to the blood upon the snow.

×**H. j. 'Preziosa'** is the most interesting of the new Hortensias, like 'Impératrice Eugénie' but larger-flowered. The Hortensia-type head opens a greenish-lilac colour but soon turns to a good deep red in sunlight. The leaves and slender stems are reddish.

H. j. 'Rosalba' (Van Houtte), A.M., R.H.S., 1939. This attractive variety was figured in the *Botanical Register* (30, t. 61 [1844]) and is quite commonly grown in Britain. The flower-head is a flat corymb with the fertile central flowers pink or blue and the six or seven marginal ray-flowers, usually four-sepalled, open white and soon turn crimson in parts, giving an effect of blotched crimson and white. With this variety a fertile flower often forms the centre of the ray-flower. The leaves are of an unusually bloomy matt yellowish-green and the habit is upright, the bush reaching 3 or 4 feet in height. It is a very pretty little Hydrangea and ideal for small shady gardens.

×**H. j. 'Intermedia'.** This variety is *H. Thunbergii* × *H. japonica*. As a clone this Hydrangea is very commonly found in nurseries and is unusually vigorous and easily grown. Its decorative value, however, is not very high compared with other varieties, owing to the fact that the ray-flowers twist over and face the ground so soon.* The ray-flowers are few in number, often only three or four per corymb. They are markedly dished in shape, with three, sometimes four, noticeably serrated sepals. Perhaps 'wavy' or 'deckle-edged' would describe them better as the teeth of the serrations are attractively rounded. The ray-flowers open white, then turn crimson and, all too soon, twist over on their pedicels until they face the ground. The stems are reddish and slender and the leaves much puckered, and reddish-green, with red veins. In poor soil and full exposure it is a weedy little thing a foot or so high; in shade and rich soil it will grow over 6 feet high. Exhibited

---

* As soon as the central fertile flowers are pollinated, the ray-flowers, their purpose of insect attraction having been completed, twist out of the way so as to assist seed ripening.

26 × *Hydrangea japonica* var. *intermedia*

by Messrs. Baker of Codsall at the R.H.S. in 1931 it received an Award of Merit, unfortunately under the name of *H. s.* var. *acuminata*. This mistake as to name has since been corrected in the records. This variety was figured in the *Journal*, *R.H.S.*, Nov., 1940 (Vol. LXV, pt. 11, t. 104). It is a useful plant for filling up around Rhododendrons in woodland, as almost any shoot stuck into the ground *in situ* in such places usually roots naturally and forms a plant without further trouble. It is surpassed by many other varieties for garden effect. (See Illus. 26.)

H. j. 'Grayswood', A.M. 1948. This variety is greatly superior to the two preceding, though of similar flower type.

The world's first ornamental landscape gardens were made in Japan, many centuries ago, and the Hydrangeas of the *japonica* group were among their first inhabitants.* Since that time, through the years, the Japanese have bred and selected them, so that it is small wonder that the antecedents of such varieties as this are lost in the dim past.

The bush reaches about 6 feet in time. The habit, at first upright, becomes bushy as the branches bend gracefully over with the weight of the flowers and strong shoots from the base take their place to form the centre. The handsome leaves are matt surfaced and rather a yellowish-green edged with reddish-brown, elliptic and tapered to a sharp point. The flower-head, roughly domed in shape, has a densely packed ring of about nine large ray-flowers surrounding the central, roughly circular, area filled with the pink or blue fertile flowers. The ray-flowers have four sepals, the lower one characteristically much enlarged and elongated and having graceful, rounded serrations on either side just above the tip. They open white, sometimes faintly blushed with pink or blue, and turn to a remarkably vivid crimson under the action of sunlight. This colour is often held until the first frost, making an established specimen of 'Grayswood' a shrub of high decorative value in the garden. As with other varieties of this shade-loving section, the first flowers usually open in early July and the plant is in beauty until winter sets in. Young plants do not flower so freely at first as other varieties. (See Illus. 16.) This variety requires care.

H. j. 'Pubescens'. This variety is really only of botanical interest, for a collection. The leaves are large and downy but the ray-flowers around each corymb are often only three or four in number. In appearance they resemble those of *H. japonica*.

×**H. serrata prolifera** (*H. serrata* var. *stellata* Wilson). This curious variety with a dense head of sterile flowers with doubled sepals was introduced by Maximowicz to Petrograd

* von Siebold, *Flora Japonica*, **1**: 106–7 (1840).

in the time of the Czars in 1864. Only one plant, it is said, survived. From this, stock was raised and it soon reached France and Britain. It is still, however, very scarce, if not extinct, in this country, indeed, I have been so far unable to secure the true plant. Specimens reported here always proved to be *H. involucrata* var. *hortensis*.

This Hydrangea was figured in Siebold and Zuccarini's *Flora Japonica* as *H. stellata* and is an ancient Japanese garden variety represented by one of Thunberg's original specimens (for the latter see Illus. 3.)

A further examination of Thunberg's original specimens, which I have confirmed since having obtained living wild plants from Japan, seems to indicate that the specimen shown on the right in Illus. 2 is *Hydrangea acuminata* while that shown at the left of Illus. 2 is the quite distinct *Hydrangea japonica*.

HYDRANGEA ACUMINATA GARDEN VARIETIES

I take the name *Hydrangea acuminata* to comprise the notably hardy dwarf woodland species that flowers early with blue flowers (pink on limy soil) on small shrubs of slighter build, rather than the massive maritime species or the red-stemmed *japonica* types with their white flowers turning red where the sunlight strikes.

'Bluebird' (R.H.S. Award of Merit 1960) shows hybrid vigour and makes quite a large bush up to 5 feet. The entire-sepalled pale blue flowers open in early July, as a rule, and as autumn approaches they turn upside down and pink in colour. As the leaves also turn red it is thus a decorative piece of garden furniture right into November. (See Plate H.)

'Diadem' (R.H.S. Award of Merit 1963) is a seedling of the pure species, a little shrub of 2½ feet with vivid blue flowers (pink on limy soil) of charming refinement of shape, in June. In sunny places the leaves turn deep red and so show up the flowers very dramatically. 'Diadem' is notably hardy, succeeding well in gardens too cold for the Hortensias.

'Miranda' is another selected seedling form with a particularly limpid blue flower and a strong bushy habit.

'Blue Deckle' was bred with the aim of producing a dwarf rival to 'Vibraye', flowering at 2 feet instead of 5 feet, and with a lighter, Lacecap, flower-head which would not get bowed to the ground by heavy rain. We have delayed putting up this variety for an Award of Merit until we have an ample stock of young plants ready to meet demand.

Chapter 5

The breeding of Hydrangeas

In the foregoing pages it has been shown how the Hortensias were originally evolved in France from the mixed parentage of *H. maritima* var. 'Joseph Banks' and × *H. macrophylla* vars. 'Otaksa', 'Rosea', 'Mariesii' and 'Veitchii' and also *H. japonica*. It has been shown how the Japanese clonal varieties of *H. macrophylla* used as foundations already contained the 'blood' of the three woodland species *H. japonica*, *H. acuminata* and *H. Thunbergii*.

It may not be unprofitable to summarise the more important breed characteristics handed on to the hybrids by the infusion of the 'blood' of the various primary species. Briefly the characteristics of *H. maritima* were:

(1) Massive stems.

(2) Shy-flowering except near the sea (fls. only from terminal bud).

(3) Huge corymbs (about eight small to medium-sized ray-flowers in the var. 'Joseph Banks').

(4) Ability to grow in fullest exposure to sun and wind, particularly sea-wind, and in limy soil.

(5) Pale flower colour – pink or blue.

(6) Only moderate *bud* hardiness, but hardy branches.

(7) Late flowering.

The characteristics of *H. japonica* were:

(1) Tall, slender stems.

(2) Free-flowering (flowers from side-shoots also).

(3) Few (four) but large individual ray-flowers with serrated sepals.

(4) A preference for shady conditions.

(5) Flower colour white, turning vivid crimson – never blue, by the action of sunlight.

(6) Only moderate hardiness, young wood rather tender.
(7) Mid-season flowering.

The characteristics of *H. acuminata* were:
(1) Tall, slender stems.
(2) Free-flowering (flowers from side-shoots also).
(3) Fairly numerous (six) ray-flowers with even-sized ovate and entire sepals.
(4) A preference for shady conditions.
(5) Flower colour pale pink or, very readily in acid soil, a pure pale blue.
(6) Extra hardiness.
(7) Very early flowering.

The characteristics of *H. Thunbergii* were:
(1) Very short, slender stems, dwarf growth.
(2) Free-flowering to a certain degree (from upper side-shoots).
(3) Very numerous (up to twelve) small, cupped ray-flowers with orbicular and entire or notched sepals.
(4) A preference for shady conditions.
(5) A rich flower colour.
(6) Moderate hardiness.
(7) Early flowering.

We have shown how the qualities of these species are strangely often still coupled, even in hybrid Hortensias which are the result of innumerable intricate crossings. Only occasionally have a few varieties appeared having a modicum of the virtues of all the species involved – the large corymbs on stout stems able to stand exposure from *H. maritima*, the large, bold deckle-edged sepals of *H. japonica*, the ready, pure blue and the free-flowering and hardy qualities of *H. acuminata* and the moderate size, vivid intensity of colour and quantity of florets of *H. Thunbergii*. The genes are already so mixed up that anything can happen, but it is very important, in breeding, to have the objective clearly in mind.

In general, there are two especially obvious objectives. One is to produce a still better market variety for sale as a pot-plant for decoration purposes. The other is to produce a hardy outdoor flowering shrub for garden use.

To provide a better market variety we must produce one that is superior to 'Foch' which is, I think, in the opinion of most growers, still the best all-round variety for this purpose. Let us consider, then, how this fine old sort could be improved. Its free-flowering qualities and flower colour leave little to be desired and, in any event, its seedling, 'Ami Pasquier', is as brilliant a pure crimson as could be reasonably expected. None the less, an entirely new colour break – such as a yellow – could provide an additional attraction, though not, perhaps, a superseding variety.

A yellow Hydrangea is genetically a possibility – even a probability. This is owing to the quality of phyllody (or leafiness) of the sepals. Already, in 'Vulcain', 'Edelweiss' and 'Papillon' and in the freak form of 'Goliath' – 'Greenmantle' – shown at the 1949 Chelsea Show by the writer, we have this characteristic developed to an unusual degree. Thus the yellow-green, xanthophyll pigments of leaves are available. Latent, they await the work of the breeder to give them the opportunity of birth. It will be remembered that it was in precisely this way that the Auricula, evolved from pink-flowered species, produced the beautiful yellow-flowered strains so well known today.★

When grown blue, 'Foch' is a deep blue; there is no comparably effective variety with a readily attained pale blue flower. 'Fisher's Silverblue', 'Lemenhof', 'G. Kuhnert', 'Holstein' and 'Paris' are fancied by many growers, but from the market point of view they have their disadvantages. 'Fisher's Silverblue' is not a very strong grower, 'Lemenhof' has too slender stems, 'G. Kuhnert' is too shy, 'Holstein' is very good but not stout enough, and 'Paris' is insufficiently free-flowering. There is, then, a possible opening for a market variety with paler blue flowers. The colour of 'Lemenhof', which matches Butterfly Blue No. 645 in the *Horticultural*

★ 'The Development of the Auricula', by the late Professor Sir Rowland Biffen, *Journal of the Royal Horticultural Society*, **67**: 187 [1942].

Colour Chart appears to be the truest and most lasting of this hue. On the whole, however, growers prefer a deeper colour, as much of the natural vividness is unavoidably lost in forcing.

Now, as to improvement in habit of growth and flower shape, there is no doubt that most growers would prefer shorter, stouter stems and a still larger and more globular flower-head with larger individual florets. These qualities would have to be present in any effective rival to 'Foch' in its own colour range of clean rose-pink, rich purple or deep blue.

It is, however, probable that, as in the case of colour, an entirely new break in flower shape would have a big success. A huge corymb composed of fewer but very much larger flowers with serrated sepals is perfectly possible. Deriving from *H. japonica* 'blood', such flowers have, with the exception of 'Holstein', usually carried weak colourings reluctant to blue. 'Beauté Vendômoise', with individual flowers nearly 3 inches across was an example. It was precisely by crossing 'La Marne', a variety with such large, serrated flowers, with 'Maréchal Foch' that Monsieur Kluis, no doubt with that end in view, produced the fine variey 'Kluis Superba'. To achieve a real break in individual flower size, however, it would no doubt be necessary to breed back to large-flowered sorts again – even perhaps to a nearly pure *H. japonica* such as the Lacecap variety *H. j. macrosepala.* Monsieur Henri Cayeux, the eminent French breeder of Le Havre, informs me that by crossing the hybrid Lacecap clone *Veitchii* with Hortensias of deep colouring he obtained 'curious plants with flat flower-heads with ray-flowers of enormous size'. It is at all events from this direction – that of *H. japonica* or new species – that any great increase in flower size must come.

The provision of stouter stems to hold up the large heads, without staking, must derive from *H. maritima*. The difficulty has been to achieve these without the less desirable characteristic of that species – a shy-flowering nature. As breeding continues, however, the genes get more and more mixed up, so that sooner or later a variety with stems as strong as those of 'Joseph Banks' but without the tip-flowering habit is quite likely to appear.

Another profitable target for breeders might be the production of an improved white-flowered variety. 'Mme E. Mouillère', the leading white-flowered Hortensia, is a glorious plant but, due to its *H. japonica* strain, the flowers tend to turn pink as they age. Some object to this and it might be possible to get a white-flowered variety free from this trait by using *H. maritima*. 'Joseph Banks' is pale pink but the wild prototype is a paler pink or even an almost pure white. Thus one might expect white-flowered offspring to appear and these should not carry the 'blushing' quality.

To sum up the objectives, then, for a new market variety its colouring would have to improve upon that of 'Foch', 'Ami Pasquier' or 'Mme E. Mouillère', its corymb be larger and more globular and carried on a stouter stem and it would have to equal their trouble-free healthy growth, their outstanding freedom of flower, and their willingness to be forced into flower when required.

The continental breeders of Hortensias are still actively producing them and many fine new seedlings are being grown-on at the present time. The production of new Hortensias, therefore, does not appear to be a field which British growers can advantageously enter at the present time.

Indeed, in my view, the breeding of the old type of Hortensia has nearly reached stagnation. For forty years × *H. macrophylla* varieties have been bred one to another, and in that time almost every possible combination of the available genes has been given birth.

What is really wanted is a new break altogether, and this could surely be achieved by an outcross to one of the fine new species within the subsection Petalanthe. Subsequent breeding should yield quite new and improved types which would lead to an enormous increase of public interest in Hydrangeas and probably a highly rewarding clear start for the successful breeder.

There will be less difficulty in obtaining some of the best species now that China is reopened. There are apparently quite a number of species of the subsection Petalanthe having larger, more vividly coloured, or more numerous flowers than the original parents of × *H. macrophylla*. We know to what heights of size and colour the flowers of that race have been brought. But surely even these would be excelled by their own offspring with the bold purple-flowered *H. macrophylla* subsp. *Chungii*, the huge-flowered *H. scandens* subsp. *chinensis* forma *yunnanensis* or *H. japonica macrosepala*, or the amazingly free-flowering forms of *H. scandens* with corymbs issuing from every leaf axil of the arching branches, like a Berberis. Hardiness hardly matters with the pot-plants: they spend most of their lives in the shelter of frames and houses anyway, but there is no reason to suppose that the new species would prove more tender.

It is a pity that so little was done to get these Hydrangeas in the years when China was more or less accessible. We could have spared a few new Rhododendron species of no particular merit in exchange for *H. scandens* subsp. *chinensis* forma *yunnanensis*! Apparently it is common enough in the Salwin-Mekong Divide.

With regard to the breeders' second objective, a hardy outdoor shrub, the situation is altogether different for the British breeder. Continental breeders have told me that they have often produced seedlings with a large and beautifully formed Lacecap type of flower-head with deep, rich-coloured flowers. They have destroyed them, however, because their object was purely to produce florists' varieties with still larger globose heads. Unlike British growers, they have only a very limited market for outdoor flowering shrubs. Thus, in this field, the way is wide open for British breeders to produce a hardy outdoor flowering shrub with the lighter, more shapely and natural-looking, Lacecap type of flower-head which is storm-proof.

To provide a better outdoor shrub than the best that is available today should not be at all difficult. The most important colour for a Hydrangea is surely blue – for no plant of any kind – even the Delphinium, the Meconopsis or the Gentian – can produce such solid masses of the purest hues of that rare and beautiful colour. The best blue outdoor Hydrangeas of that informal Lacecap type are the new varieties 'Diadem' and

'Blue Deckle' of *H. acuminata* and the variety 'Bluebird' and 'Bluewave' (a chance seedling of × *H. m.* 'Mariesii' brought out in 1904). Both the latter sorts have obvious drawbacks for ordinary garden use; they require shade and their colouring is weak in comparison with that of the most vivid of the Hortensias. Their hardiness, though adequate for the majority of our British gardens, is not quite sufficient for general planting throughout the colder districts of the country. There appears to be no reason why an absolutely hardy, vivid blue Hydrangea with a shapely and natural-looking flower-head fit to take its place among the elect of hardy outdoor flowering shrubs should not be attainable by suitable hybrid breeding.

The hardiest and best garden plant among the other Hydrangea subsections would appear to be *H. paniculata.* Its hardiness is perfectly adequate for all parts of the British Isles. In the chapter dealing with the species, the pink-flowered hybrid already bred by Monsieur Foucard between *H. paniculata* and a deep-coloured Hortensia is described.

This cross could possibly be made again and no doubt it would be necessary to select a Hortensia variety of a ready blueing variety and to breed back to this to obtain the required intensity of colour. It should be mentioned, however, that the Hortensia head is dominant and the Lacecap recessive.

I should state here that there is an incompatibility between the species of Hydrangea of the different subsections but this is not fully understood. That it is not an insuperable obstacle is proved by the hybrid mentioned above and also by the hybrid bred by Monsieur Cayeux between × *H. macrophylla* and *H. anomala* subsp. *petiolaris* of which particulars are given in the chapter devoted to Species.

Monsieur Louis Mouillère informs me that in 1938 and 1939 he found the incompatibility between *H. paniculata* and a deep-coloured Hortensia variety too great. The hybrid seedlings obtained showed the characters of both parents clearly, but were not strong enough to live long. Furthermore, when these species were grafted on to each other the grafts only held for a very short time.

In 1935 Messrs. Mouillère also crossed *H. paniculata* with *H.*

heteromalla forma *Bretschneideri*. Seedlings were obtained which, while retaining the paniculate flower-head of *paniculata*, had the orbicular sepals of *Bretschneideri* on the sterile flowers. The latter were only normally numerous – not preponderant as in the case of *H. paniculata* var. 'Grandiflora'. A small red 'eye' decorated the centre of each sterile flower, but a superior form of the wild type of *H. paniculata* (*H. p.* 'Floribunda') shares this charming characteristic.

By courtesy of the Royal Horticultural Society, Dr Janaki Ammal was so kind as to carry out some chromosome counts of *H. maritima*, *H. acuminata*, *H. Thunbergii*, *H. japonica* and *H. m.* 'Joseph Banks'. She informed me that they were all diploids, 2n = 36. The others counted are also assessed at 36 as follows: *H. cinerea*, *H. anomala* subsp. *petiolaris*, *H. quercifolia*, *H. arborescens*, *H. arborescens* subsp. *radiata*, *H.. heteromalla* forma *xanthoneura*, *H. scandens* and *H. paniculata*. *H. paniculata* var. *floribunda* has 72 and so has *H. p.* var. *praecox*! *H. aspera* was tested by E. Schœnagel in 1931 and found to have 36.

If the difficult cross, × *H. macrophylla* × *paniculata* could be made again, it would probably provide a quick solution of the problem, but crossings with other species within the subsection Petalanthe would be a comparatively easy matter. Unfortunately there seems little hope of obtaining the most desirable species for the purpose and so we must make the best of what we have available.

Even without the extra hardiness obtained by hybridising with hardy species an improved garden flowering shrub could probably be obtained by merely breeding from carefully selected varieties of × *H. macrophylla*. Some varieties are notably so much hardier than others that they habitually carry their wood quite undamaged through normal winters; others always suffer severe damage. Those varieties with a strong strain of *H. acuminata* such as *H. m.* var. *rosea* Veitch and its offspring 'Vibraye', and also certain other old varieties with evident 'hybrid vigour' such as 'Bluewave', 'Whitewave' and *lilacina* are examples of unusually hardy sorts. 'Bluewave' × 'Souvenir du Président Doumer' or 'Westfalen' are crosses that naturally suggest themselves.

Mr. J. M. Grant, head gardener at the famous gardens of Grayswood Hill, Haslemere, sowed seeds of 'Bluewave' and also of 'Grayswood' about 1936. The seedlings of both these hybrids showed, predominantly, *H. japonica* characteristics. This seemed to indicate the strength of this strain in the composition of both the parent varieties. On the other hand, I have found that the seeds of *H. japonica* always germinate more readily than those of the other foundation species and the seedlings thrive better.

We have ourselves recently flowered a large batch of seedlings of 'Whitewave' and these show great variations, there being Hortensia heads like 'Joseph Banks', deckle-edged *japonica*-type flowers and *Thunbergii* types, in fact all the ancestral species, but all show pale colourings. Consequently, to secure an improved Lacecap variety, it seems that it would be necessary to select a really deep-coloured Hortensia variety for one parent, and, probably, to cross this with a flat-headed Hydrangea with flowers that come blue unusually readily, such as *H. acuminata*.

There is still some mystery attached to the origins of the foundation clonal variety 'Otaksa'.★ I believe that there is some tenderness in the strain which accounts for the fact that many Hortensias are not even as hardy as *H. maritima* var. 'Joseph Banks', *H. acuminata*, *H. japonica* or *H. Thunbergii*. The strong, permanent, woody growth made by 'Joseph Banks' would be a big asset to any Hydrangea bred for use as an outdoor shrub. This plant is also, therefore, one to bear in mind in breeding for the new objective.

An intermediate type of flower-head retaining the lightness and grace of the Lacecaps, but with more vivid colouring and an extra row of sterile flowers would probably provide the ideal outdoor shrub.

To achieve a blue-flowered shrub the species *H. japonica* and all strains showing its influence would have to be avoided. Thus 'Whitewave', 'Mme E. Mouillère', 'La Marne', 'Deutschland' and all its group comprising 'Hamburg', 'Peer Gynt', 'Elmar', etc., could not be expected to

★ See page 61.

produce good blue-flowered offspring. On the other hand, the readily blue-flowering character of *H. acuminata* is very marked.

In the past, due to its not being fully realised whence the various characteristics were derived, Hydrangea breeders have often been working in the dark, having to rely upon trial and error. By using the most suitable hybrid varieties having the required species characteristics dominant it should be possible to plan the breeding more confidently.

As regards the Cornidias, a fascinating and entirely new field lies open to the breeder. The rewards of producing a vivid-flowered evergreen climber would be high, even if the plant required a coldhouse in the colder districts. *H. Oerstedii* should be easily imported (see pp. 77, 78) from South America and there are wonderful plants among the other species. Nowadays there are a great many unheated glasshouses, and glassed-in verandahs where Camellias are grown. The Cornidias should prove invaluable plants for late summer flower in the shadier parts of such places, or outdoors in the south and all up the west coast.

Among other Hydrangea species of possible value to the hybridist may be mentioned the best form of *H. aspera* forma *villosa*. The quality of greatest value is its readiness to give blue flowers, even on alkaline soil. On the other hand, this species is not a very easy one to grow and in my experience is very often difficult to start. So far as I know, no experiments have yet been made to test its compatibility with *H. macrophylla* or with *H. paniculata*.

A point that might be mentioned is the desirability of giving more thought to the naming of new varieties. A Hortensia variety is a world-wide plant and its name should be internationally practical. Such names as 'Petite Sœur Thérèse de l'enfant Jésus', 'Gloire de Boissy-St. Leger' and 'Gartenbaudirektor Kuhnert' are not practical. On the other hand, such names as 'Mousseline', 'Paris', 'Rosebelle', 'Swanhild' and 'Amaranthe' are not only practical but charming and descriptive as well. Brevity is very desirable, considering the thousands of times the name must be pencilled on a label if the variety is to retain its identity and, consequently, attest its breeder's renown.

POLLINATION AND SEED RAISING

With the Lacecap section of × *H. macrophylla* seed is readily obtained, but care is necessary to secure seed from the Hortensias. It will be noticed that each little subsidiary stalk of the corymb usually carries one or two fertile flowers in addition to the sterile ones, and these can be recognised at an early stage in the opening of the corymb. Some varieties – such as 'Vulcain' – mature very few fertile flowers, others such as 'Foch' and 'Altona' have numerous, strong fertile flowers. To assist in getting good seed some of the adjacent sterile flowers may be snipped off at an early stage.

An essential in all planned breeding is to prevent the female parent flower being pollinated by its own pollen. Consequently the anthers, or pollen-bearing organs, of the seed parent must be removed before they can fertilise the adjacent stigma. In addition, precautions must be taken to prevent any insect agency transferring pollen to the selected stigma. The fertile Hydrangea flower being small, the operation of removing the anthers requires small sharp scissors, good eyesight and a steady hand. It is, however, unnecessary to prepare more than a very few flowers, as the seeds are so tiny that one or two capsules will contain sufficient for raising several hundred plants.

The next operation is the application of the pollen of the desired male parent to the stigma when this is in a viscid and receptive condition. Some varieties produce little or no good pollen, others produce plenty, and some produce a pollen which appears to be toxic.* It is helpful to pollen production to remove the sterile flowers adjacent to the fertile flowers, but great care has to be exercised or too many exposed wounds

* According to Mr. P. F. Morris of Melbourne.

may cause the corymb to fail. Strong sunlight may burn the exposed fertile flowers and, on the other hand, too much shade may prevent their full development and ripening.

Another complication is that there is a wide difference in the times when the varieties mature their sexual organs. The pollen, unless the male parent variety is brought forward, may not be ready at the right moment. Pollen that is ripe too soon can, of course, be stored for some time in special containers but it is generally less troublesome to bring on or retard the plants. In view of the uncertainties, the wise breeder will have a good selection of varieties prepared beforehand and a cool lath-house and a heated greenhouse convenient for his operations.

Assuming that the fertile flowers are duly fertilised, and seed set, there ensues a long period of waiting before the seed is ripe. Normally it will be about the middle of December before the capsules are ready. The minute seeds are generally of a brown colour and as tiny as those of the smaller Rhododendrons. I have had the best results from sowing these at the first promise of spring in pans whose drainage holes are covered first by perforated zinc, then by a layer of broken crocks and finally by sifted, sterilised John Innes compost. In my own experience the seedlings come up more thickly and vigorously in this compost than in the peat and sand compost recommended by many French breeders. No doubt there is a difference in the mineral content of that valuable but highly variable substance, the peat, that accounts for this. The seeds are sown thinly on top of the thoroughly firmed and levelled surface of the compost, and the pan is covered with a pane of glass and kept warm but lightly shaded. For watering, the pans are partly immersed in warmed water.

As soon as the young plants are large enough to handle with a tiny pen-like trowel, made from a short length of Bamboo cane, they are pricked-off into boxes of more acid and peaty compost with sharp drainage and grown-on as fast as possible, being kept warm, moist and lightly shaded. When they are 4 inches high the seedlings are put into small pots and potted-on in the usual manner.

If the rapid growth described by Monsieur Cayeux (p. 186) has been achieved, the seedlings may be forced and flowered within little more than a twelve-month as he describes. Otherwise they may flower as late as their second or third year, but, in any event, the waiting period for the first flowers is trifling compared with that to which Rhododendron breeders are accustomed. When the breeder produces a winner, there is little difficulty in propagating a good stock of it from cuttings in a short time and thus the rewards of success are more immediate.

Chapter 6

The cultivation of the garden varieties

Hydrangeas are not to be recommended for people who dislike gardening and require what are called 'trouble-free' plants. The Hydrangeas are dependent on our care; they are immediately responsive to our skill and especially reward those who enjoy the craft of growing fine plants, rather than merely owning plants that have grown themselves. Even in the smallest and humblest garden, a sturdy Hydrangea at once proclaims the place as the abode of a real gardener rather than a mere purchaser of plants to cover the ground.

To grow really well outdoors, the Hydrangeas require either naturally suitable conditions, with an acid, loamy soil in a reasonably mild district, or specially arranged soil conditions and also the most favourable garden climate that can be provided by aspect and air-drainage.

The ideal climate for them is one in which either nearness to the sea-coast or relative elevation ensures comparative freedom from early autumn and late spring frosts.

As I travel about the country, always on the look-out for Hydrangeas, I note, invariably, splendid specimens on all the hills where air-drainage is good, and no plants or only poor ones in the frost-holes and low-lying lands.

This is because the Hydrangea shares with other first-class shrubs, such as the Rose and the Tree Peony, that fatally sanguine disposition which leads them to go on making growth too late in the season and also to open their buds too early, often at the first mild spell in late winter. Apart from this failing, the Hydrangea is quite hardy enough for open-air culture in most parts of Britain.

In elevated gardens free from untimely frosts, the Hydrangeas make great permanent bushes yards across and high, provided that they are not endangered by untimely pruning.

In such favoured places they may be grown in light shade so that their delicate flower colours are preserved from being bleached by the sun. In lower-lying places they must be grown in poorer soil and in *full sun and exposure* so that they are encouraged to ripen the season's wood fully against winter, and spring pruning may be necessary.

One of the first principles of good cultivation for a fine specimen flowering shrub is that annual feeding should be carried out as a regular routine. This principle is clearly understood and practised in America and on the Continent but, strangely enough, this is not so in Britain. As a result, having been carefully planted in good soil, the shrub at first grows well and may even become a fine specimen but, in forming this framework and providing several heavy crops of flower, it exhausts the nutriment of the surrounding soil. Consequently no sooner has it reached its prime than it begins to go back. This is perfectly avoidable. Every year a mulch of leaves, bracken and fertiliser should be given to replace the wastage. In addition the plant's own spent flowers and leaves should be left. Nature's way is to drop these sources of plant food from the branches that no longer need them on to the roots below that are ready to process them, so that their contents may be used again in the future. Thus the mulch is an easy and effective natural way of feeding a plant. Consider, then, the enormous amount of petalage produced by a good flowering shrub and be generous in replacing the wastage every year. In dry, hot summers, liquid feeding may also be necessary.

Foliar feeding with liquid solutions absorbed by the leaves, such as F.F., is often very effective.

SOIL AND FLOWER COLOUR

As to soil, the Hydrangea, like most fine flowering shrubs, prefers a definitely *acid* soil, particularly a loamy, lime-free soil fairly rich in humus (or decayed vegetable matter). A turf-loam to which a little rotted leaf-mould or peat has been added is ideal. A lime soil, which is of course the opposite, is bad because *such a nature makes many mineral plant foods unavailable to plants*. Iron and aluminium are two such foods, required by the Hydrangeas, that they cannot get out of a limy soil. We can, of course, feed these substances to the plants artificially but it is a troublesome business, except, of course, in the case of those grown in pots or tubs.

The soil requirements of the hybrid Hydrangeas are pretty definite. Too limy a soil causes chlorosis – yellow leaves lacking the green chlorophyll – due to iron starvation. This occurs because a plant growing in limy soil lacks the power to assimilate the iron and other trace elements like aluminum and possibly because these elements are made more difficult to assimilate by the alkaline soil conditions. It is possible to reduce the alkalinity of soil by applying about 3 lbs. of powdered sulphur and 5 lbs. of aluminum sulphate per hundred square feet and rotavating this in so thoroughly that it is completely incorporated in the upper 18 inches of soil. I have done this successfully in cases where Azaleas and Hydrangeas were demanded where the soil had previously been artificially limed. It would be helpful if more research could be done on the subject, with exact measurements and scientifically controlled experiments. Gardeners often make the mistake of thinking that the baled peat from deep excavations is acid and so will acidify limy soil. Unfortunately it has no such effect. It is only the substitution of large quantities of surface peat dug from acid-soiled woods or bracken land that can bring about the desired result. Accelerated compost is limy and so are fallen leaves from limy areas.

The acidity of soil is measured by a scale called the pH scale. On this scale pH 7 is neutral, pH 6·50 is just acid, pH 5·00 admirably acid from the Hydrangea's point of view and pH 8 excessively limy. Now, the health of a Hydrangea plant depends largely on its getting plenty of iron which makes the leaves, which, by photosynthesis, nourish the plant, healthy and of a deep green. In addition the vigour of the plant is assisted by getting supplies of aluminum. There is plenty of both these elements available in most acid soils.

For years no one knew why the flowers of Hydrangeas were blue in some soils and pink in others. Many, myself included, thought that iron had something to do with it. Mr. E. M. Chenery of the Agricultural Chemistry Department, Imperial College of Science and Technology, London, gave us the first inkling as to what the causes really were. In the *Journal of the Royal Horticultural Society*, **7**: 304 [1937], he gave an account of all the early experiments and his conclusion that aluminum was the cause of blue flowers. He recommended 1½ oz. of aluminum sulphate being placed on the surface of each 6-inch pot *during the dormant season*. The proportion required being equal to 2½ lbs. of the crystals per cwt. of compost.

Later the whole matter was exhaustively gone into, with innumerable carefully conducted tests, at the Department of Horticulture and Ornamental Horticulture, Cornell University, Ithaca, N.Y. The results, given by Mr R. C. Allen,★ clear up the whole position in an admirable manner.

Briefly, the aluminum, sucked up by the plant in large quantities, acts upon the crimson pigment (Delphinidin 3-glycoside) in the sepals of the flowers, so that they become bright blue. Thus if you see a glorious great Hydrangea bush covered with brilliant blue flowers you may be sure that it is growing in an acid soil or is being fed with iron and aluminum.

The blueness of the flowers depends exactly on the acidity in accordance with the table below which is a condensed example of one given in the work cited:

Soil pH 4·56 intense blue flowers, aluminum content of the flowers 2,375 parts per million.

" pH 5·13 blue flowers, aluminum content of the flowers 897 parts per million.

" pH 5·50 mauvish flowers, aluminum content of the flowers 338 parts per million.

" pH 6·51 mauvish-pink flowers, aluminum content of the flowers 214 parts per million.

★ *Contributions to the Boyce Thompson Institute* by R. C. Allen, **13**: 221–42.

" pH 6·89 pink flowers, aluminum content of the flowers 180 parts per million.

" pH 7·36 clear, warmer pink flowers, aluminum content of the flowers 100 parts per million.

Obviously, then, anybody keen on growing really good Hydrangeas would be well advised to get his County Horti-cultural Department (financed by the ratepayers) to make a series of pH tests of the soils in his garden.★ He will then know just where he is and whether or not he needs to feed extra iron and also, if he insists upon blue flowers, aluminum to his plants. If he finds that his soil is limy (or alkaline) he may decide to go in only for the varieties with white, rich pink or crimson flowers. In this case he will feed his plants with a solution of ¼ oz. of chelated iron to the gallon of water whenever he observes the symptoms of iron starvation. This is evidenced by what is called chlorosis, a yellowing of the foliage and stunted growth consequent on the leaves not having enough chlorophyll to function properly.

It should be mentioned that the figures given above were obtained with the variety 'Souvenir de Mme Chautard'. In the open ground this is not a ready and reliable bluer like 'Vibraye' or 'Bluebird'. For example, in our typical slightly acid soil of pH 5·50 some of Chautard's flowers are a mauvish-blue, some pink and some pale blue. In the same soil 'Vibraye', 'Foch', 'Bluebird', 'Mousseline' and many others are a uniform and perfect pure blue. 'Rosea', 'Bluewave', 'Parsifal', 'Goliath', 'Blue Prince', 'Kluis Superba' and 'Altona' are a purplish-blue in their early years unless given a little aluminium feeding. The German group of 'Hamburg', etc., and also 'La Marne', 'Elmar' and 'R. F. Felton' are lilac. On the other hand, the red varieties, 'Ami Pasquier', 'Carmen', 'Heinrich Seidel', 'Gloire de Vendôme', 'Vulcain' and 'Westfalen' require an occasional watering with a solution of ¼ oz. of lime to the gallon of water if they are required to retain their pure red flowers and not to show quite attractive, but sometimes unwanted, violet hues as

★ This service is also carried out, for a small fee, by Farall Nurseries, Blackdown, Haslemere. They do not carry out soil analyses which are a different matter and quite unnecessary for the purpose.

they mature. 'Princess Beatrix' usually remains red at this degree of acidity.

Feeding with liquid manure, made as usual from farmyard dung, is helpful during active growth in the early part of the season if the plants are not making stout enough shoots. A useful tip, to prevent any unpleasantness arising from the liquid manure tank, is to add a handful of sulphate of iron occasionally.

Now, you may ask, how is the gardener with a neutral or only slightly acid soil to achieve bright blue Hydrangea flowers? The answer is that he must feed the plants with the aluminum which enables them to provide flowers of the proper colour. This may be done in various ways. Perhaps the least laborious method is to lift the plant carefully in autumn immediately after leaf-fall and then dig out the soil of the part of the bed where it is growing, putting this on to the adjacent path a barrow-load at a time. With this soil 2 lbs. of aluminum sulphate is thoroughly mixed. The soil is then replaced and another barrow-load may be similarly treated.

This plan works quite well provided that winter rains are normally copious and that there are no other mineral deficiencies. On the other hand, if an abnormally dry winter is followed by a dry summer, as in 1948, the plants die owing to the unusually small amount of moisture, making the solution too strong. Root action is evidently discouraged by excess aluminum and it retards the growth of young plants. Consequently I prefer to grow these on as fast as possible, watering with liquid manure and only ¼ oz. of sulphate of iron and a mere pinch of aluminum sulphate per gallon of water when necessary. When the plant is well grown and established, liquid mineral feeding, as described below, will ensure blue flowers if the colour requires improvement. Thus, rather than forcing the plant unnaturally, we work in harmony with its natural tendency by withholding extra mineral feeding until its second season.

The best method is to build up a rim of soil around the plant, as it grows, so that it appears to be in the centre of a saucer whose rim comes just outside the widest spread of the branches. Then a solution of the same type of aluminum at ¼ oz. to the gallon of water and ¼ oz. of sulphate of iron may be given in autumn and again, when growth starts in spring. About 2 gallons of this solution may be fed to the plants every week in spring until the flowers are fully open. Thus it is a slower and more tedious method in the long run. Even then the operation must be done with care and thoroughness so that beyond all doubt *all* the soil in contact with the roots is saturated by the solution. It may even be necessary to pierce the root-ball with a slender pointed stick to get full penetration. If the solution is made too strong the foliage is destroyed, the leaves turning brown and dying. Even so, new growth is usually soon made and the plants seldom die outright from this cause, provided, of course, that the error is not an exaggerated one. The iron should always be added at the same time, as it helps in assimilation.

In the course of blueing, the first part of the inflorescence to respond is that composed of the fertile true flowers. In the Hortensias these are inconspicuous, one only being usually found on each stalk bearing four or five sterile flowers. With the Lacecaps, on the other hand, the fertile flowers are almost as conspicuous as the sterile ray-flowers and thus their colouring is important. White Lacecap varieties should therefore receive, either by the natural acidity of the soil or by artificial feeding, enough aluminum to ensure the blueness of these if the plants are to show at their best.

Another method which gives little trouble, but is much more risky to the health of the plants, is to apply on the surface of the soil, all over the area occupied by the plants' roots, a dressing of aluminum sulphate in November, after leaf fall, and again in spring before growth starts. For large bushes about 10 lbs. *in all* would be required. Thus, at 20p a lb. it would cost about £2.00 per plant by this method. Mr. Chenery observes that with smaller plants, about ¼ lb. of aluminum sulphate per stem may be taken as a rough guide as to the quantity required.

It is necessary, I find, to mix the aluminum very thoroughly with the surface soil and to make sure that, if the weather is not suitable, gentle regular showers of water (with ¼ oz. of iron sulphate per gallon), per the watering can, take the place of rain to diffuse the substance gradually among the roots. If the winter is very dry, and heavy showers only come after growth has started, the plants are apt to be killed by the sudden excessive dosage they receive.

In the course of my work, I have tested the pH value, or acidity, of a great many soils. I think it is probable that the greater number of typical, natural, acid soils found in this country rate at about pH 6·00. This acidity is not quite sufficient for the majority of Hydrangea varieties to achieve a pure blue without added aluminum sulphate and iron. The readiest bluers will have good blue flowers in time, but without help they may take a few years to come true in colour. In such soils some feeding with both iron and aluminum is of help.

All blue-flowered Hydrangeas are liable to revert to pink flowers when moved, and cuttings of even the purest blue-flowered plants if struck and grown in insufficiently acid soils always give pink flowers. The necessary aluminum is stored in the leaves and flower-heads rather than the stems, consequently, when it is necessary to move a plant, even from one part of an acid-soiled garden to another, it is wise to add a little aluminium sulphate to the soil in the new position. This should, however, be kept well away from the roots as these will penetrate the untreated soil more actively.

Certain varieties, having the genes of the blue-flowered species dominant, readily absorb the aluminum and thus readily give blue flowers. Others, having more of *H. japonica* in their composition, have less facility and are therefore troublesome if blue flowers are required. Examples of particularly ready 'bluers' are 'Générale Vicomtesse de Vibraye', 'Gartbr. Kuhnert', 'Mousseline', 'Maréchal Foch' and 'Mme Riverain'. Obdurate varieties that do not readily give blue flowers are 'Liebling' ('My Darling'), 'Elmar', 'Rochambeau', 'Beatrix', 'R. F. Felton' and many, such as the German group like 'Hamburg', whose serrated sepals patently betray an infusion of *H. japonica* blood.

It may be mentioned that should it be desired to change a blue-flowered Hydrangea to pink flowers this is only too readily done by adding 2 lbs. of ground limestone to each barrow-load of the soil in which the plant is grown or by watering with lime solution as advised. It will take a little time to exhaust the supplies of aluminum already stored in the plant, of course, but the final result is certain enough.

I have dealt with the flower colouring of Hydrangeas in some detail because I have often been astonished at the misconceptions current even among quite experienced growers. I have listened to salesmen telling prospective customers that 'Hydrangeas are not naturally blue. They have to be treated with chemicals.' Actually in their native land of Japan or in any of the normally acid soils *predominant in these islands* the real blue sorts are naturally blue.

Nature always seeks to provide an acid soil and this, not limy soil, is the normal fertile soil for vegetation. If we were to board an aeroplane and take a trip to the rain-forest, the most fertile soil on Earth, with our soil testing outfit, we should find that the prodigious growth was made in a soil of very high acidity. When we speak of applying such fertilisers as nitrogen, potash or phosphate to other crops we do not call it 'treating them with chemicals'. In acid soil it might equally be said that it is necessary to 'treat the plants with chemicals', *e.g.*, apply a lime solution, to attain pink or red flowers.

ASPECT

There is another important factor in the cultivation of Hydrangeas and that is the question of aspect; whether we should grow them in full sun or in shade. I am afraid that the answer to this question is a complicated one. The wild maritime species of the Hydrangea, *H. maritima*, grows always, as I have said, in the fullest exposure, but it, and its hybrids, flower, our Japanese friends tell us, 'in the overcast and rainy month of

July'. Thus the delicate colours are preserved when the flowers open, but the plant gets full sun and air to make and ripen its growth. In this country, then, unless we grow our Hydrangea in full sun and wind and can then provide a canopy of cloud over it the moment the flowers open, there is bound to be some conflict of interests. Are we to risk soft, tender growth in shade so as to preserve the flower colours, or grow the plant in full exposure and risk the hot sun of an English July burning away the flower colours? In practice there are several ways of getting round the difficulty.

Hydrangeas in tubs, of course, are easy. We can leave them in the sun until the flowers are opening and then wheel them into the shady courtyard or round to the north side of the house. Actually, so long as there is clear sky overhead, exposure to direct sunlight is not absolutely necessary for firm growth, except in very cold gardens, and therefore the bed at the foot of the north wall of the house is a particularly good place to grow Hydrangeas in districts where sun-heat is intense. They get enough exposure to ripen their wood well, and the flowers get sufficient shade to prevent them getting bleached.

There are other loopholes. In Cornwall and up the west coast the sun, if I may be forgiven for saying so, seldom shines for long periods on end, and there the Hydrangeas grow and flower, beautifully tinted, in that superb, mineral-rich, acid soil, right out in the open. Then, again, we may grow the plants with clear sky overhead, but distant trees so placed that they keep the midday sun off the flowers. Finally, there are certain varieties whose blue colour does not fade very much unless the sun is abnormally hot, and to grow these is really the best of all ways out of the difficulty. These kinds are described in the chapter on the varieties.

The aspect required for a given variety depends much on the primary species which it favours (see p. 142). *H. maritima* types usually enjoy sun, *H. japonica* types hate the sun, *H. acuminata* and *H. Thunbergii* types are not very particular.

The British climate varies so widely from year to year that it is impossible to lay down an infallible ruling. In a freak summer with a burning sun for weeks in midsummer, such as that of 1973, many plants in the sunnier positions will flag dangerously and require regular watering. This may be facilitated by standing a 12-inch pot close alongside each plant. The pot is forced firmly down a few inches into the soil and then a trowelful of soil is thrown in. It will then be found that the plant sucks the moisture slowly and steadily so that a filling will last for at least a couple of days. Even this is a troublesome business, but less so, on the whole, than the replacement of shaded plants whose insufficiently ripened growth has led to their death in winter.

Furthermore, it will be found that plants in the shade of trees, sufficiently near them for the tree-roots to reach the soil in which the Hydrangeas are growing, will suffer more disastrously in a prolonged drought than plants in the open in deeply dug and well made beds kept properly mulched.

Unusual sun heat will entail the destruction of some of the leaves by sunstroke, but, if these are snipped off as soon as they are seen to be dead, the loss is not usually a serious one.

The Hydrangea is a very thirsty plant, and, in order to avoid excessive watering work, if the soil is not naturally moist we shall need to alter matters so that it becomes so. This requires to be done in two ways. In the first place we have to make sure that the upper layer of the soil is kept moist and this is easily achieved by keeping it covered with a thick mulch of bracken, fallen leaves, hop manure or some such, or even at a pinch, lawn mowings.

Next we have to ensure moisture at the lower levels, for the deep roots. Here I will say that I have measured the roots of an old Hydrangea bush and found them to go down for over 4 feet! Very few soils often get dry at that depth, but we have to cater for the plant's first few years, whilst the roots are on their way down. So, to do this, we need to work the soil deeply and put in, deep down, some juicy material such as rotted dung. From this sponge, moisture finds its way up to the roots above. We may also bury sandstone rocks, or lead a soakaway from the roof gutters of the house to the bed or take any other measures that may occur to us. As a rule the place selected for the Hydrangeas will be moist enough at the deeper levels, but on some dry soils the above precautions have to be taken to ensure success. As well as in all other conceivable situations, I grow Hydrangeas on a hot south slope of pure white sand. On that slope, in different parts, I took all the precautions mentioned and they all seem to work quite well.

FERTILISERS

As regards fertilisers for Hydrangeas, they are not difficult plants in this respect and profit by small doses of all the usual ones. Bracken seems to have some special virtue that encourages free rooting and is so notably appreciated that I have come to prefer it to anything else. An essential point is that the bracken fronds shall have been cut in June, when they are almost, but not completely, unfolded as they are then full of plant foods. After this, the food materials are carried down to the rhizomes (fleshy roots) for storage until the next growing season. Thus the rhizomes themselves, if dug up in winter and killed by drying in the following summer, provide a very rich plant food. Such a rhizome, buried near a Hydrangea plant, will be almost immediately penetrated by a root in the most convincing manner. I think that the reason for this remarkable effectiveness as a plant food is that the chemicals are already selected, refined, diluted and mixed in the right proportions by the bracken plant. Thus the Hydrangea is provided with a perfectly prepared dish ready for immediate consumption, so to speak, instead of merely the crude raw materials.

Various mineral plant foods affect the colour of the flowers, as I have pointed out, but there is another way in which this may be done which I have not so far mentioned. Many of the newer varieties, in particular 'Vulcain', 'Heinrich Seidel', 'Ami Pasquier', 'Arthur Billard', 'Rochambeau', 'Westfalen', 'Président Doumer' and others have beautifully coloured crimson, carmine, or red flowers and gardeners on neutral soils may well be fascinated by these and wish to attain the maximum vividness of hue that is possible.

To attain this a neutral turf-loam containing plenty of fibre and applications of dried blood are helpful. But to get the full warmth of hue attainable it is essential to make certain that the pH value of the soil is really pH 7. The flower colour of Hydrangeas is even more affected by the acidity than by the variety. An interesting experiment which will convince any-one of this may be made as follows. Secure twelve rooted cuttings of, say, 'Maréchal Foch'. Take a barrow-load of soil known to be very acid and one of a soil known to be very limy. By gradual admixture arrange twelve potfuls covering full gradation from the very acid to the very limy. Plant a cutting in each and observe the flower colours. If the gradation has been carefully arranged, the tints will run from warm pink through magenta and purple to pure deep blue.

PRUNING

Now as to pruning, generally speaking the hard pruning of established Hydrangeas does them more harm than good and may easily be fatal. If the older wood is cut away after flowering, a mass of soft young shoots is produced which will be winter-killed almost certainly. Thus a fatal cycle of annual soft growth, killed each winter, is set up, and spring pruning is essential to stop it.

At all times the pruning of any flowering shrub should only be done to achieve a certain definite object. Young Brooms, for instance, must be kept sheared down to rounded, domed mounds of dense green twigs, beautiful in the winter garden landscape and alight with massed flowers in early summer. Roses need to have time-expired stems removed. The Hydrangea is naturally of rounded, dense habit in a properly open position. Its size, varying from a foot to 8 feet both ways according to variety, is ideal for any given space if proper selection of variety is made. Why, then, should we prune? The objectives are to ensure the freer flowering of shy-flowering varieties and to assist frost-damaged plants to recover quickly.

Some gardeners admire the persisting flower-heads of the Hortensias and value them as part of their winter garden decoration; others object to them and wish to prune them away in autumn. This is a rather damaging operation. In the first place it is not for nothing that the Hydrangea has learned and achieved the wise practice of providing for itself a winter mantle of dry flower-heads to protect the buds below from frost. Secondly, the flower-heads contain relatively large quantities of aluminum in easily reassimilable form, and to rob our soil of the constituent that enables the flowers to achieve their beautiful blue colour is obviously a most unwise proceeding.

With the Lacecap varieties, however, the spent flowers may be snipped off to prevent the plant exhausting itself by ripening a heavy crop of seeds. The cut should be carefully made just above the second pair of leaves from the top.

In general the best pruning is merely to cut out the three oldest stems *at ground level* in late winter. A young Hydrangea shoot bears a flower at its tip the first year. Next year two side-shoots are made and they flower – and so on and so on. When the shoot has made fifteen or twenty flowers in this way it is time to remove it from the base to make more room for younger shoots. Needless to say, cutting young shoots half-way down merely prevents the shrub flowering. Indeed, no pruning at all of any flowering shrub is better than indiscriminate hacking. The only ones that really need regular pruning, not counting the Roses, are *Spartium junceum* and garden varieties of Broom to keep the bushes to a dense, low, mounded habit and Wistarias to remove the unwanted whip shoots and turn these into short, flowering spurs.

SUMMER PRUNING

Some modern varieties notable for very large corymbs and having a strong strain of the maritime species in their composition are not naturally free-flowering. Unlike those whose ancestry contains a stronger infusion of the woodland species, they do not flower from side shoots if the terminal bud is winter-killed. More regrettable still, the terminal bud, even, does not always flower. With these kinds we may carry out summer pruning in an attempt to improve matters as recommended by Marcel Ebel in the French handbook mentioned on p. 173. Immediately after flowering the inflorescence is removed, with a short length of its stalk, so that the remaining part of the stem will 'plump up' the buds left so that they are more likely to produce flowers. More important still, this pruning will ripen the wood more perfectly against winter. The less free-flowering varieties, of which 'La Marne', 'Monsieur Ghys', 'Oriental', 'Mme Legoux' and 'Paris' are among the finer examples, are seldom really successful as outdoor shrubs in any but favourable gardens. Such plants should, however, profit by their advantages and give some of these massive-flowered sorts a trial.

SPRING PRUNING

After an unusually severe winter or in very frosty gardens, spring pruning may also be necessary for all varieties to ensure getting at least a few flowers from the damaged plants, and, above all, to prevent the loss of the new growth in the following winter. Some of the branches will be completely dead; others, though they may seem at first sight to be in much the same state, are still alive except for their tips. The latter should be carefully pruned off to a point just above a joint on live wood. Flowering shoots will spring from the joint, provided that the variety is one of the free-flowering ones and that the plant's energies are not permitted to expend themselves on sending up too thick a sheaf of young shoots from the base. These basal shoots must be thinned out to about half their number as soon as they appear, leaving just enough to form the framework of the new bush that will have to be grown. The few shoots left, thus given full light and air and the nourishment that would have gone to the many, have then a better chance of ripening to a fit state to stand the next winter.

Once a Hydrangea of one of the hardier varieties has made a well-furnished bush it is practically safe from ordinary frost damage for life, provided that it is not interfered with. Ancient bushes in every county in England testify to this. Once autumn-pruned, however, unless care is taken to spring-prune, also, so as to make good the damage caused, the plant seldom re-establishes itself properly.

Unsuitable varieties for outdoor culture, on the other hand, can only be made worthwhile in favourable gardens where they also receive regular pruning attention. The double-starred varieties, however (see list), are so free-flowering that an inch or two of surviving old wood is enough to provide four or five corymbs and new shoots from the ground will also flower in the same season. But even with these, every effort should be made to get them to build up into stout, twiggy bushes that require no pruning at all.

In late summer there is one little attention that comes, perhaps, under the heading of pruning, that is well worth while. Certain varieties, such as 'Altona', 'Blue Prince', 'Hamburg', 'Deutschland', 'Brightness', 'Mousseline', 'Flambard', 'Princess Beatrix' and others have the characteristic of holding their flower-heads in perfect condition long after the original colouring has faded and then, without losing succulence, been replaced with delightful green, red or purple autumnal tints. These are most decorative in the autumn garden picture, but a fungus of some sort sometimes attacks and spoils these flowers. One or two browned florets, perhaps, may be found in an otherwise attractive head. Unless these florets are at once snipped off, carefully and individually with scissors, the whole flower-head soon becomes infected and spoiled. On the other hand, if the infection is thus promptly removed the head will probably remain in beauty for months on end.

The above notes on cultivation refer principally to the Hortensia section of × *H. macrophylla*. The Lacecap section require, on the whole, similar treatment, but their hardier nature, and larger infusion of woodland 'blood', enables us to grow them satisfactorily in shadier places. In their case we shall often need to do this, because, unlike the Hortensias with their 'garden variety' look, the Lacecaps associate delightfully with Rhododendron species and other natural-looking types of flowering shrubs grown in wild gardens and wood gardens.

MOVING HYDRANGEAS

All but the very largest, long-established Hydrangeas can be moved at almost any time of year as safely as Rhododendrons, and by the same method. This is to take a ball of soil sufficiently large to surround the roots adequately. It is mainly a mechanical problem, requiring skill and care, and above all gentleness, in the case of the larger specimens.

Young plants, perhaps a couple of feet across, may even be moved from one part of the garden to another when in full bloom. The ground is given a good soaking the previous evening and the surrounding soil is gently removed, starting at a sufficient distance from the plant to ensure that the ball of soil encloses all the roots. The actual lifting is done with two large forks and a piece of burlap may be used to carry the plant to a point near the new position. The forks are again used to place the plant in the amply wide hole made for its reception. Great care is taken to make certain that the soil level, after replanting, is exactly correct, as before. If very hot weather follows, a piece of burlap, supported on canes, may be used to keep off the full glare of the sun.

When pot-bound plants are put out in the open ground it is advisable to loosen the lower part of the ball of roots. Otherwise the plant may fail to start its roots working out into the new soil. Growth remains inactive, and watering only aggravates the evil. In such cases it is always best to dig the plant up again carefully, open up the root system gently, and replant. But, in my experience, pot-grown plants put up a very poor performance compared with those specially grown in the open ground for planting out, as the latter have a wide and powerful root-system of quite different character.

In a very loose soil containing much humus and fallen leaves, the Hydrangea makes masses of surface roots instead of deep roots down into the subsoil. These surface roots get insufficient moisture in a dry summer so that poor flowers, chlorosis and insufficiently ripened wood result.

The remedy is to brush aside the mulch from all around the plant and replant it, burying the surface roots in firm soil. The mulch may be replaced in late spring by which time the deep roots will have made further growth downward and thus ensured the continuous moisture required for the health of the plant.

There are two new chemicals that help us in growing Hydrangeas. Before moving a bush in summer we can spray the leaves thoroughly with the anti-transpirant S.600. The leaves then will not wilt until the covering is washed off, so that if care is taken to water only at soil level quite astonishingly successful moves can be made.

The other useful new product is Benlate, the systemic fungicide. This enters the plant's system helping it to resist fungus infection. Thus it is useful for 'Altona', 'Westfalen' and other long-lasting flowers as it helps to preserve the flowers from botrytis.

Chapter 7

Propagation, forcing and pot-culture

I know no shrub of comparable value that is so easily prop-agated as the Hydrangea. The tip of any half-ripe shoot carrying also two pairs of leaves and cleanly severed, will usually root and form a new plant in any reasonably moist, sandy, lime-free soil so long as direct sunlight is kept off sufficiently to prevent its wilting at the start.

Even an old live stick of Hydrangea wood, stuck into the ground, perhaps to hold a tie-on label for some other plant, will often send out roots.

For the amateur, a simple box, half filled with moist sand, or a mixture of sand and rotted leaf-mould, covered with a close-fitting glass lid, will provide for his modest needs with an almost 100 per cent take of cuttings. Layering, carried out in the usual manner, is a reliable and easy method of providing a small number of strong young plants.

For the professional, many different systems are in use to provide the larger output required. One leading British prop-agator of the choicer Hortensia varieties simply inserts the cuttings in the natural soil under trees or in a special type of lath-house formed of hop-netting on wire and post supports with corrugated iron sheets at the sides to keep off ground draughts. Others strike their cuttings in pots of pure sand placed in a warm greenhouse or use propagating houses or frames.

The great French propagators often proceed as follows.★ Ground for frames is prepared by preliminary digging to loosen the soil and then about 8 inches depth of sand with a little peat mixed in, is *added*. The finished surface is arranged at a distance of about 15 cm. from the glass of the lights. Thus raised, the cuttings-beds are warmer and there is no danger of

★ *Hydrangea et Hortensia*, by Marcel Ebel. J. B. Baillière et Fils, Paris.

stagnant moisture. This work is done a fortnight before the cuttings are introduced so as to give time for any weed seeds present to germinate and then be destroyed by the hoe. Between April 15th and May 15th the Hortensias put out into the open ground the previous year and destined for sale in the autumn are in full growth; the time is ripe to cut them back, even if the cuttings are not required.

If this system (see p. 180) is not followed the early cuttings required may be obtained from old plants brought on in warmth for the purpose.

The operation of removing the shoots is done in the early morning when they are well plumped-up from the freshness of the night air.

Large baskets are used, lined with wet cloths, and the cuttings are placed horizontally in these. Only those that can be inserted during the day are taken, as otherwise such soft shoots become flaccid and are unreliable.

Before the cuttings are inserted, the baskets are submerged in water for an hour. Each cutting is about 3½ inches long. This usually provides a joint at the base and one in the centre from which the leaves are cut off and the tip, which is left intact. When, at about the end of ten days, the young cuttings are rooted, they are pinched back. Some days later, when new growth shows, air is given so that the lights may be dispensed with as soon as possible.

The great danger, with this method, is damping off and to avoid this some growers remove the lights the moment the cuttings are rooted and pinch them back later. This plan has much to recommend it.

When a keen and able staff is available to ventilate the frames at every opportunity, closing down at the first hint of cold and protecting them further with bracken or litter in severe spells, damping off can be prevented. Otherwise the menace is such that any method which will lessen the risk is worth considering.★

In the case of scarce varieties, cuttings are sometimes made of small pieces so as to obtain a greater number of plants from

★ Spraying with Benlate seems to be a promising system.

the available material.† These pieces may be only single joints with a small piece of internodal stem below and even these may be split down the centre, after rooting, so as to form two plants. Thus each eye produces a plant; warmth, ventilation, and care and skill are, of course, essential, as such pieces are very vulnerable to fungus attack.

GROWING-ON IN THE OPEN GROUND

In France and southern England the young rooted cuttings are often planted out in the open ground and freely watered until they are re-established. A mulch is also applied to conserve the moisture. About a fortnight later, the cuttings are pinched back and this is repeated about mid-July. By August, in a favourable season, the young plants will have about a dozen or more branches.

During the growing period the plants are frequently watered and dressed with sulphate of ammonia. If the foliage is not naturally of a deep healthy green, sulphate of iron is also applied. This may be added to the water given, making a solution at the strength of ¼ oz. to the gallon. Liquid manure may also be given with the iron solution and this has the effect of deodorising the manure water completely. At the end of August feeding is stopped, and in September the plants are lifted and potted-up.

Alternatively, another system may be followed, the young plants being pinched back only once and merely mulched and not watered at all except in extreme cases. By this means the plants will be hard and their growth should be sufficiently ripened to stand the winter outdoors. To protect them from early frosts, stout posts and wire-netting with string-netting, reeds, heather or bracken laid on top are used. In France this affords sufficient protection.

† 'Boutures de Tailles avec Bourgeon Axillaire', by Henri Cayeux. *Revue Horticole*, Suisse, August, 1936.

In the spring, when growth recommences, the plants are pruned down to two good 'eyes' and the ground is hoed and top-dressed with guano or other fertilisers. The young plants are kept growing vigorously and pinched back again in mid-June. In July, feeding is stopped so that the growth may ripen off. In August, potting takes place, the plants being extra fine specimens with about twenty or more branches.*

The risk of severe frost damage to *young* plants in the open is undeniable, but in climatically favourably situated gardens, where there is little skilled labour available, it will often be found that young plants wintered outdoors beneath a thick mulch of bracken and with some overhead protection suffer fewer casualties than those under glass, and grow away more quickly in spring. This is because a natural leaf-fall and ensuing dormant period are essential to the vigour of the plants.

GROWING-ON IN POTS

Hydrangeas do not really take very kindly to pots, having an extensive and eager root system they suffer quickly from the least neglect. It is only by constant pinching, feeding and repotting that they are kept actively growing.

The plants thus grown fetch a higher price than those from the open ground. They have more flower buds and are more highly finished. None the less the extra cost of the work entailed makes their production, in France, but little more remunerative than that of the open-ground plants. In these times, more and more Hydrangeas are grown on in the latter manner and the practice is now very frequently followed in Holland and the south of England also.

With plants in pots, watering is the biggest job and care should be taken to organise this conveniently, for a plant suffering the smallest neglect in this regard quickly goes back and is unsaleable when required. There must be no such check

* *Hydrangea et Hortensia*, by Marcel Ebel. Baillière et Fils, Paris.

to growth. By pinching, feeding and repotting the plants must be saleable by autumn. About three repottings are usually done. After each pinching the plants are repotted and the pots plunged in rows, taking care that they are not so deeply buried that the roots escape into the surrounding soil over the tops. Each week, liquid feeding is given, manure water to which a little sulphate of iron is added being very effective.

SOILS AND COMPOSTS

As regards potting composts, any good standard type of peat, leaf-mould, loam and sharp sand mixture in the proportion of three parts of loam to one of humus will serve, but for blue flowers it must be acid. A pH value of about 5 is ideal. When there is any doubt, and blue flowers are required, about 2 lbs. of aluminum sulphate per barrow-load of compost should be thoroughly mixed in. Care should be taken to use soft water – a hard water, containing lime, prevents the plants assimilating the necessary iron and aluminum for pure blue flowers.

When crimson-flowered varieties are grown and required to show this colour rather than purple or dark blue, a neutral compost is required. One very slightly on the acid side of neutral, say about 6·85 can, however, be used. Iron solution, made by dissolving ¼ oz. of sulphate of iron in each gallon of water should be given if the plants appear unthrifty, with pale or yellowish foliage. Dried blood should also be used as a fertiliser. This has an appreciable effect on the richness of the colouring.

Established growers come to use standard compost ingredients whose correction to the exact required acidity for the colouring desired has become a matter of practice and experience. But even then they may be caught napping by an unexpected change in the acidity of one of the ingredients. When this happens a fine batch of one of the pure crimson-flowered varieties may come with flower colours of an unattractive puce tint or, if the change is towards alkalinity, the choicest blue sorts may show nothing but mauvish-pink flowers. *Thus a pH test of the composts and their ingredients each year is a very desirable safeguard.* This may easily be carried out by the grower himself with one of the inexpensive field soil-testing outfits sold for the purpose by British Drug Houses, or the samples may be sent to Messrs. Boots (Horticultural Dept.) for them to carry out the pH test.

Sometimes it is only found at a late stage that the flower colours are not going to be sufficiently pure and vivid owing to insufficient care in composting. If the undesirable colour is spotted in time, by noting the hue of the upper stems (showing a reddish or a bluish cast) or perhaps by an odd, extra-early plant showing flower, something may yet be done to improve matters, by adding either aluminum sulphate for the blues, or lime for the reds and pinks, to the water used for watering. But it is a troublesome and risky business, great care having to be used in the exact measuring of the strength of the solutions so as to avoid a check to growth at this delicate stage.

BRITISH PRACTICE

In England more protection generally has to be given than is necessary in France. A leading firm of growers, who produce many thousands of remarkably fine Hydrangeas each year, use the following method. The stock plants are kept in boxes wintered in freely ventilated frames. About a week before Christmas the tops are cut off and the plants brought into growth. In response to the warmth provided, innumerable shoots push out and these are taken, as cuttings, into a heated propagating house where they are dibbled into a peat and sand mixture in close rows. When rooted, they are transferred to boxes and pinched. Then, later, they are planted out to grow on, pinched again, potted, to ripen off in late July, and sold, for forcing, in the autumn. Thus well-branched, strong young plants are secured that may be relied on to respond freely to forcing later. By the following April such plants are well bloomed, with from five to nine flower-heads, depending on the variety, and in May may be seen decorating window-boxes, stands at race-meetings, hotels, etc., in their full glories of bright pink, sky blue, lavender, crimson, purple or cobalt.

OUTDOOR HYDRANGEAS

In our own nurseries the object is totally different. The Hydrangeas grown are required to provide as strong and hardy a young plant as possible for planting out in the open, in spring, as a permanent outdoor flowering shrub. To secure this, the plants are grown entirely in the open air and are never pinched, but encouraged to make one or two very stout, fully ripened, stems only. Thus, if the top is killed after planting out, the strong dormant buds safely below ground push out ample new growth later. Had these been previously forced into action by pinching, the plant's recovery would be much more uncertain if the whole of the top growth were lost.

The young plants are grown 'hard' and wintered outdoors, protected only by a mulch of bracken. In spring, as soon as the danger of severe frost is thought to be over, taking into consideration the unusually favourable situation of the nurseries with regard to katabatics, the young plants are lifted, with a ball of soil to retain the large root-system, and sorted. Thus, any tendency to making precocious growth is checked and the plants are kept dormant. In this state they are packed and despatched.

FORCING

In France, M. Marcel Ebel tell us, Hydrangeas may even be forced so as to secure saleable plants in bloom as early as Christmas. In this case operations should begin in October. Given suitably prepared material, *that has been well ripened off in the previous summer*, about two months are required from the start to the time the flowers are sufficiently open for sale. But this is not easily done as, lacking a proper period of rest, the foliage is apt to have a tired look. Unlike Azaleas, Roses, etc., which are finished after forcing, Hydrangeas actually improve. Even backward plants flower well later on and, being larger, are improved in value. This is because the Hortensia has a natural tendency to flower at the earliest possible moment. Thus, instead of resisting, the plants co-operate freely. Forcing may be done in various ways.

The temperature may be gradually raised to 20°C., and later to 22°. As soon as the corymbs are sufficiently developed the temperature must be gently allowed to fall to about 15° or malformed flowers may result.

Feeding and the provision of the necessary minerals for proper colouring is carried on as previously described, and as soon as the forcing is considered sufficiently advanced, the temperature may be allowed to fall to 10° or slightly lower.

BRINGING ON

In France and parts of the south of England Hydrangeas in frames or cold houses come into growth naturally in early April. At that time heat may be given for a fortnight to provide a temperature of 15°, care being taken to protect the plants from sudden keen frosts. After this, more ventilation is given, aiming at an average temperature of about 10°. By late May or early June, provided that the plants were well prepared beforehand, they are ready for sale. To attain this some trimming may be necessary. Weak growths that cannot flower and also strong basal flowerless shoots are removed. Fumigation or spraying may become necessary to destroy aphis or other pests.

VARIETIES FOR FORCING

In France the most popular varieties for forcing, from November, for the early trade, are 'Maréchal Foch', 'Ami Pasquier', 'Etincelant', 'Rouget de Lisle', 'Mme E. Mouillère', 'Merveille', 'Louis Sauvage', 'Rosebelle', and 'Splendens', but it is interesting to note that, according to Monsieur Marcel Ebel, many very old varieties of particular value for outdoor planting are still the choice of many growers in France for forcing today. 'Générale Vicomtesse de Vibraye' is an example.

Monsieur Ebel gives 'Maréchal Foch', 'Mme. E. Mouillère', 'Mme F. Travouillon', 'Monsieur Ghys', 'La France', 'Hollandia', 'Oriental', 'Paris', 'Goliath', 'Mme Legoux' and 'Eclaireur' as the choice of French growers for forcing in the second period – from December 15th to January 15th.

For the third forcing period, which is merely 'bringing on' rather than true forcing, much the same varieties find favour.

I have thought it of interest to give these particulars of French production owing to the much larger scale on which French growers operate. A large number of firms have, each, as many as a hundred thousand plants on hand at one time.

Mr. G. Hargreaves of Stockbridge House, EHS (*Gard. Chron. and Horticultural Trade Journal*, Vol. 174, No. 19, p.24) recommends the following varieties:

Blue, early:	'Deutschland', 'Enziandom', 'Holstein' and 'Lemenhof'.
Pink, early:	'Pink Princess', 'Prima'.
Mid-season:	'Benelux', 'Europa', 'Gerda Steiniger', 'Harry's

Pink Topper', 'La France', 'Mrs. W. J. Hepburn', 'Rosita'.

Red, mid-season: 'Harry's Red', 'Morgan Rood' ('Morgenrodt'?).

Late: 'Succès', 'Eldorado', 'Mrs. R. F. Felton', 'Rheinland'.

Many general nurseries grow a few Hydrangeas each year. They cannot afford the space and labour required to produce superfine specimens of all varieties. In such cases the advantages of certain remarkably easy and responsible varieties of quite exceptional freedom of flower may be considered. Such are 'Princess Beatrix' for pink, 'Fisher's Silverblue' for mid-blue, 'Kluis Superba' or 'Foch' for deep blue, and 'Ami Pasquier' for red. It is true that buyers often ask for novelties, but these are not often an improvement on exceptionally good older varieties that have been selected by reason of their special merits from the hundreds bred in years past. The breeder sends out his finest products, but only the actual trade grower can tell, after several seasons' testing, whether the new variety is really an improvement on his well-tried old favourites.

Some buyers insist on having only varieties with very large individual flowers, such as 'La Marne', 'Heinrich Seidel', 'Hamburg' or 'Altona'. These are types showing the stout habit of the maritime species with, in varying degrees, its concomitant lack of freedom of flower. None the less, the individual corymbs are so large that they provide quite a fine effect.

Of the newer sorts 'Paris', a somewhat freer-flowering example of this type, is very popular and 'Kluis Superba' is favoured for a dark blue. In France 'Rouget de Lisle' still leads among the smaller growing varieties and 'Mme F. Travouillon' has a large following. 'Louis Sauvage' and 'Merveille' are also much grown for forcing. In Europe generally, on the whole, 'Europa', 'Deutschland', 'Maréchal Foch' and its red-flowered offspring 'Ami Pasquier', with 'Mme E. Mouillère' to provide the white, are probably those most commonly grown for the purpose.

In Britain 'Maréchal Foch', 'Kuhnert', 'Hamburg', 'Europa', 'Holstein', 'Merveille', 'Deutschland', 'Carmen', 'Gertrude Glahn', 'Kluis Superba', 'Altona', 'Mme E. Mouillère', 'Ami Pasquier', 'Louis Sauvage' and 'La Marne' are apparently the most commonly grown for market purposes.

Surprising as it may seem, some British growers discard the older varieties every year and grow only the latest importations. The continental breeders do their utmost to prove worthy of such extraordinary faith in their skill, but it is not every year that a variety of the outstanding excellence for market purposes of 'Kluis Superba' makes its appearance.

In many general nurseries the Hydrangeas are in confusion as to name and, indeed, few plants are more difficult to identify than the Hortensia varieties. Yet in flower colour, hardiness, habit and time of flowering the distinctions are so marked that a wrong variety is almost certainly unsatisfactory for the required purpose. The remedy is to dispose of all doubtful plants as unnamed, and to start afresh with an authentic stock comprising a limited selection of the finest varieties only. Six distinct varieties kept carefully labelled, at any rate at first, would soon become readily identifiable by the operatives.

Some growers have told me that, as names do not matter to their customers, they have asked their suppliers merely for a wide selection. This is, I think, a mistaken policy as naturally all the varieties sent are not equally good. Thus the inferior ones often remain unsold until they are finally got rid of at a low price, to clear. This adversely affects the profit on the consignment, as these flowerless inferior sorts cost just as much as the good ones.

In the British market, at present, Dutch varieties have a big advantage over French and German varieties. The Dutch actively and efficiently seek export orders to Britain and do all they can to supply well-packed, healthy, vigorous plants, clearly labelled and true to name. Naturally, and very properly too, they send their own varieties for the most part, unless French or German varieties are specifically asked for by name. British growers who have not so far tested such varieties as 'Maréchal Foch', 'Mme F. Travouillon', 'Rouget de Lisle', 'Paris', 'Etincelant', 'Merveille' and 'Westfalen' might well ask for these. I have no doubt that the Dutch exporters, being good businessmen, would arrange to supply them.★

The popular varieties mentioned are satisfactory for producing the large numbers of even-sized and even-coloured plants required, with the minimum of trouble. The displays of these exhibited at shows do, however, tend to lack variety. Sufficient attention is not always given to securing the exact pH value of the compost required to attain the desired colouring. Due to forcing, the colours are unavoidably weak and uncertain and only a pale shadow of the rich and vivid tints which these same varieties show at their normal flowering time when grown out-of-doors. By adding varieties of richer colouring, grown in soils of carefully tested pH values to suit the hue required, an enormous improvement in the diversity and richness of colour could be gained for such exhibits. Unfortunately for the market grower, 'Vulcain' which can show wonderful green and orange hues, 'Westfalen' in deep velvety purple and 'Président Doumer' and 'Arthur Billard' in intense crimson, violet or deep cobalt, are not easy varieties to force so as to get them ready in time to be at their best with the general run, unless they are started earlier. It could be done, however, and I think the public would be willing to pay the increased price that would have to be charged for these more difficult sorts.

The form of the flowers of forced Hortensias often leaves something to be desired. Blown up to monstrous size, with some varieties the very numbers of the florets make the heads mere lumps of muddled petalage. Many will disagree, but I think that the corymbs are more attractive to the eye when the form of each individual floret can be seen and appreciated. An admixture of Lacecap varieties, with their large bold ray-flowers, standing out individually would also, I think, add to the diversity and appeal of such exhibits.

★ There are, however, a few black sheep who fill up with worthless varieties, and it should be pointed out that British-grown plants are unequalled for quality and price when all import expenses are taken into account.

THE PROPAGATION OF THE WILD SPECIES

Those who require only a moderate number of cuttings of the more difficult species, will find glass bell-cloches highly effective. Owing to the natural moisture of the soil making artificial watering unnecessary, the danger of damping off is greatly minimised. A piece of mutton-cloth may be used to shade the cloches from hot sun. This is best stretched over a light frame of sticks some way above the cloches as too much shading seems to retard rooting. Small frames can also be used when a larger number of cuttings is required.

DAMPING OFF

If damping off, caused by a fungal growth, *Botrytis cinerea*, appears on young cuttings in a frame, I find that removing the frame, lights and all, replacing it with a shading material and then giving a strong application of Lugol, Benlate or Folosan or other fungicidal dust is as good a method of cure as any. Generally speaking this destructive visitation only occurs when the atmosphere is very warm and moist. Thus in lath-house propagation it is rare, unless the weather is exceptionally close.

At all times dead Hydrangea leaves should be removed from the vicinity of cuttings, as these offer a ready breeding ground for infection. The tendency nowadays is to take shorter cuttings and to trim off the tips of the leaves. This minimises the risk of their drooping and thus helps to prevent infestation.

PROPAGATION FROM SEED

The propagation of Hydrangeas by seed is an interesting method. The varieties, of course, do not come true – every seedling is a potential new clonal (or vegetatively propagated) variety.

The seeds of the Hortensia or the Lacecap varieties of × *H. macrophylla* and many other species require until about December to ripen. Each seed capsule contains over 200 seeds, as tiny as those of the alpine Rhododendrons.

Sowing is done under glass in March or April in well-crocked pots or pans filled with a sterilised compost such as 'John Innes'. The seed is thinly and evenly distributed on the firmed and levelled surface of the soil and a pane of glass is then laid over the top until the seedlings are up. If the vessels are watered by part immersion and capillary attraction, using warmed water, and a slight warmth is maintained in the house, germination should take place in about ten days.

If quick results are required the following method, described by Monsieur Henri Cayeux,* may be followed. As soon as the seedlings are large enough to handle they are pricked out and afterwards potted-on as required, keeping them growing actively all the time close to the glass. The best compost for this period, according to this famous French breeder is one containing about two-thirds of heather-peat, one-third of rotted, forest leaf-mould and some sharp sand. Pushing them on as fast as possible, by August the seedlings are in 6-inch pots and for these the soil mixture is made up with equal parts of heather-peat, rotted leaf-mould and loamy soil, with a little sharp sand added.

By September the young plants are well developed and the terminal flower bud is formed. Watering is then stopped and the plants are protected from early frost in well-ventilated frames so as to encourage a natural leaf-fall.

By wintering them 'cold' under glass, the plants can be

* 'Hortensia Plante Annuelle', by M. Henri Cayeux. *Revue Horticole*, Paris, February, 1935.

subsequently forced into growth very early and, by forcing, a good flowering can be obtained in little more than a year after sowing. To succeed with this method skill and experience are, of course, necessary and much work is entailed in maintaining the necessary warmth. Normally seedlings do not flower until their third year.

The amateur, if he has a garden with a favourable climate and soil can often succeed by sowing the more easily obtained, naturally fertilised and abundant seeds of the Lacecaps in a cold frame or even in the open ground. Indeed, in my garden these plants frequently provide strong self-sown seedlings.

The difficulty with open-ground seedlings is to ensure their safety from slugs and woodlice. Paths of sharp cinders around all seed beds, frequent dusting with insecticide and 'Meta' slugicide, always fresh and kept dry by a covering, are essential safeguards. Rows of pane cloches may be used with advantage to protect young plants of the hardier sorts over their first winter.

They may either be forced in the early spring to secure a few flowers from the more precocious specimens, or allowed to grow-on naturally, when they will usually flower in their third year.

MINIATURE HYDRANGEAS

It is possible to produce miniature Hydrangeas 2 inches high in full bloom in tiny pots for dolls' houses or tray culture. This is done by striking small adventitious flower buds as cuttings. These adventitious buds are produced by plants specially grown from cuttings of flowering wood the previous season. With care, they root freely enough, and the flowers expand reduced to a miniature size with the pair of leaflets below them, but no further growth can take place as there are no growth buds present on the cutting.

The tiny plants are very attractive and I should imagine that they would find a ready sale if such propagation was done commercially. A tray a foot long will accommodate a dozen of the little pots showing white, crimson, pale blue, dark blue, pink and purple flowers and they last in beauty for a very long time.

For the decoration of the miniature landscape gardens made upon trays, these tiny Hydrangeas are most effective though, of course, they require annual replacement.

Chapter 8

Diseases and pests

Young Hydrangeas are in danger from *Botrytis cinerea*, or 'la toile', the fungus which causes 'damping off', until they are safely hardened off for outdoor life. In fact, the plants are so much healthier free from the damp and stagnation of a glassed-in atmosphere that a number of growers in the south keep the plants outdoors from the very beginning.

Dusting with a fungicidal dust such as green sulphur dust has been recommended in the chapter on propagation as a method of combating the fungus. In France, fungicidal sprays are used to treat the frames, pots, cloches, etc., as a preventive routine measure when out of use. Should an attack of botrytis develop in spite of these precautions the affected cuttings are removed and destroyed, the surface of the sand is removed and coarse common salt is freely applied. Some of the remaining healthy cuttings may be destroyed but the attack is controlled.* Benlate spray is better still.

Frost damage is not, of course, a disease, but the treatment required for a damaged plant may be mentioned here. It is essential to remove all dead leaves at once as these form a ready breeding-ground for fungus. Dead wood should be removed by a clean cut into the sound wood below, just above a joint.

Aphides are always liable to become troublesome and cause the curling up and destruction of the foliage. The moment these pests are noticed, a spraying with any of the proprietary insecticides recommended for the purpose will usually put an end to them for the time being. A further spraying a week or so later is, however, advisable as survivors soon breed up again and fresh airborne infestations may arrive.

In wet places slugs are very destructive to the young shoots of Hydrangeas below and at ground level. At times they also

* Monsieur Marcel Ebel.

climb up the stems and eat their way into the pith. A watch should be kept for these and one of the 'Meta' type slug poisons, arranged in small heaps, will soon kill the vermin off. Young plants may also be given the protection of a perforated zinc collar arranged to encircle the stems within a barrier about 4 inches high and firmly pressed into the ground.

One may wonder what happens eventually to the thousands of Hydrangea plants produced and sold each year. Some go into the dustbin, but I believe that many are planted-out in gardens after their flowering is ended. During the ensuing winter the soft growth of the top is killed back to ground level. In the following spring the portion below ground often sends up some young shoots, but the moment that they appear the slugs mow them off. Thus the plant bleeds to death and dies, and the gardener, seeing no growth, believes that the plant succumbed to frost. To prevent this, failing the provision of a zinc collar, a shovelful of sharp weathered ashes over the crown at planting time and repeated applications of slug poison in spring are very necessary precautions.

Cockchafer grubs or 'Joe Bassetts', cutworms and wireworms are all to be reckoned with, but a really strongly growing mature Hydrangea bush can stand a lot of nibbling without noticing it very much. Good cultivation and fertilising and a watchful eye are needed, however, or one may easily lose a fine bush owing to a sudden attack by migrating slugs.

Compared with such regrettably palatable plants as Roses, Crabs, Cherries, Clematis and evergreen Azaleas the Hydrangeas are remarkably free from caterpillar pests.

In my experience rabbits will not touch Hydrangeas, but I am told that on limy soils chlorotic plants are sometimes attacked. Perhaps the quantity of iron and aluminum in the foliage of a healthy plant makes it distasteful.

A form of mildew is very destructive to the foliage of Hydrangeas in glass-houses or frames, but I have not seen plants grown entirely outdoors as flowering shrubs attacked. Hydrangea mildew usually appears in late summer and is curable by spraying with a sulphur fungicide wash, such as

Sulsol, or by fumigation with sulphur candles where this is feasible. Here, again, Benlate may be the answer.

The old variety 'Souvenir de Claire' was much addicted to mildew and it was to avoid this, Monsieur Louis Mouillère tells me, that the cross with the strong and more or less immune variety 'Whitewave' was made. Many admirable varieties, such as 'Maréchal Foch', 'Holstein' and 'Kluis Superba' seem slightly less resistant to mildew than others. The disease is readily controlled, but immediate treatment is essential or all foliage may be destroyed.

Red Spider is a pest that often attacks young plants in frames or houses. The leaves will be found to lose colour and vigour and a careful examination will reveal the minute red specks on the undersurface that are responsible for the mischief. Spraying, at once, with an insecticide such as Chlorocide, is advisable or growth will be severely checked. I have not seen Hydrangeas grown entirely as outdoor shrubs attacked by red spider, though in prolonged dry spells young plants in the open from frames may suffer.

The more favourable the micro-climate the less Hydrangeas suffer from pests and diseases. In no garden is there a place with a micro-climate so favourable as that at the foot of the house walls. For, with all our care in insulation, an occupied house gives off such a lot of warmth as to have a decisive effect on the plants growing at the foot of the walls. Houses are now such warmth-givers that, with the decrease of air pollution, London has as favourable a climate as Cornwall!

Indeed, Geraniums that I planted in Mayfair survived last winter outdoors whilst those in a friend's garden in Marseilles were killed by the cold! So do not let builders ruin the wall-foot with concrete or paving – let the house rise out of flowers and keep paths a yard away from the walls.

Chapter 9

Hydrangeas as garden shrubs

The cult of flowering shrubs has become ever more popular as gardeners find it increasingly difficult to provide the labour for the maintenance of herbaceous borders and the periodical planting-out and removal of tender plants bedded-out. Once properly installed and grown to fair size, flowering shrubs can provide even more brilliant and beautiful permanent flower effects with but a quarter of the maintenance work.

It is often found, however, that the tendency has been to plant too large a proportion of spring-flowering shrubs. Consequently the garden became drab and colourless from the end of May onwards. If, on the other hand, care is taken to plant species and varieties which flower at midsummer, and also those that flower in late summer there is no difficulty whatever in securing a succession of bloom to cover the whole season with a brilliance equal to that of the spring display.

Roses, freed from the totally unnecessary hideousness of the conventional 'rosebed' and planted in association with other shrubs, and, still more, Hydrangeas, are the most valuable of late-summer-flowering shrubs. Indeed, only the hybrid Azaleas can compare with them in the masses of beautiful colour which they can provide. The only snag is that, unfortunately for those whose gardens are either situated on naturally limy soil or, worse still, have had their soil ruined by misguided applications of artificial lime, both these glorious genera require an acid soil. Indeed, I would go so far as to say that the Roses also require such a soil, for on limed ground they are usually martyrs to 'black spot', and their growth lacks real vigour.

It may not be out of place to recapitulate that, in common with Rhododendrons, Azaleas, Camellias, Heaths, Roses, Brooms, Eucryphias and many other fine shrubs, the Hydrangeas suffer from mineral starvation on limed soils and, to a less extent, on chalky soils. This prevents the flowers showing the proper blue colouring of their native land and causes the foliage to suffer from chlorosis which entails a weakly and stunted growth. This condition may be improved by feeding the plants artificially with the minerals 'frozen up' by the lime. The details of such feeding are given on pp. 157–163.

The majority of gardens in the British Isles have a naturally fertile acid soil. In most cases this soil has had its fertility reduced by applications of lime. This liming habit is simply a survival of the old practice of using lime to counteract the over-manuring of kitchen gardens which otherwise adversely affected the lime-tolerating vegetables. Unfortunately it has come to be a standard recommendation of hack writers for the popular press, as though liming invariably benefited all acid soils. That this is a fallacy can readily be proved by any observant gardener prepared to test the matter out for himself.

The problem is how to restore freely available minerals to soils that have suffered from liming and the best method is probably to work-in powdered sulphur and to apply aluminium sulphate and chelated sulphate of iron to plants showing symptoms of mineral deficiency.* At first, odd-coloured flowers will betray unequal distribution of the mineral fertilisers but, on this being corrected, the plants will ultimately settle down satisfactorily.

Assuming, then, that soil conditions are reasonably satisfactory, let us consider how best we may make use of the Hydrangeas as garden shrubs. In the first place the most important position in any garden is undoubtedly the ground immediately surrounding the house. For such situations, alongside the house walls, Camellias, Roses and Hydrangeas are pre-eminent. From early spring to late autumn their beautiful flowers continuously succeed one another. The foot

---

* In America they treat limy soil wholesale with aluminum sulphate until a pH value of about 5 is attained, as a matter of course where Camellias, Rhododendrons and other choice shrubs are required to be grown.

of a south wall is rather too hot a position for Hydrangeas, but the other aspects are all suitable. Unless a red-flowered variety, such as 'Ami Pasquier' or 'Westfalen' is wanted it will be necessary to remove all the poor soil, contaminated with lime mortar and builders' rubbish, usually found in such places and to replace this with acid turf-loam. We may then plant a harmonious group composed of such varieties as 'Vibraye', 'Mousseline', 'Holstein', 'Niedersachen' and 'Amethyst' in Cambridge blue, 'Maréchal Foch', 'Altona', and 'Kluis Superba' in deep blue, and the white 'Mme Mouillère', and in the foreground the dwarf-growing 'Président Doumer' and 'Westfalen' with their deep velvety purple flowers and 'Gloire de Vendôme' in cherry-red late in the season. 'Princess Beatrix', a crimson that does not 'blue' easily in acid soil, may also be added if it is desired to add a further contrast.

From the earliest days of July until frost the Hydrangeas will be in bloom and it is just this type of long-season plant that is required for such a position as this, that is constantly before our eyes. It is the perfect substitute for bedding out.

In limy soils the red-flowered varieties such as 'Vulcain', with 'Ami Pasquier' and 'Westfalen' associated with white, and also with peach-pinks such as 'Liebling', 'Altona' or 'Violetta' will be the best choice.

In very dry districts, or in unusually hot summers, the plants may flag and wilt at times, but it will be found that they usually recover perfectly if given a good soaking in the evening. In very prolonged hot spells, such as occurred at midsummer in 1973, it is a good plan to sink a 12-inch pot an inch or so in the ground close up against the plant and to fill this with water. If the pot is pressed firmly into the soil and a trowelful of earth is put in, the liquid will only soak away very slowly so that the plant is able to take all the water it requires in its own time, and the branches soon hide the receptacle. This is a much more convenient and effective method than pouring water over the roots at intervals.

A glorious bush, forming a solid mass of exquisite colour a yard or two across, is worthy of such individual attention. Such specimens, personally cared for by the owners, are the glory of many a cottage garden, and often put to shame the extensive herbaceous borders of larger places.

For beds and borders in the more formal parts of the garden, Hortensias grown in beds in association with Roses, Rhododendrons and Genistas (such as *G. virgata*, *G. cinerea* and *G. œthnensis*) will be found highly delightful. Provided that the soil is made good before planting and a thick mulch of bracken or leaves is used to conserve moisture such fine shrubs give little trouble and improve in beauty year by year.

In low-lying places, often known as frost-holes, where the cold airs collect in early autumn and spring, Hydrangeas are not successful unless only the very hardiest varieties are selected and these are carefully spring-pruned and grown in full sun so as to ensure thorough ripening of the season's wood.

For the best colour arrangements for garden decoration it is best to make the dominant hue a 'Cambridge' blue and to diversify this with just a few 'Oxford' blues, purples and even reds and pinks at selected points. Few colours are so spectacular and effective for garden decoration as pure sky blue; in fact only pure yellow equals it. Thus in hundreds of gardens that are now dull after midsummer Hydrangea time in late summer could equal if not surpass the glories of the Azaleas in spring.

The colours of the Hortensias seldom clash; it is just a question of arranging the best contrasts. This requires some thought, because, *unless the varieties are carefully selected to suit the acidity of the actual soil*, the colours are likely to be too much alike (see p. 156). Should this initial mistake have been made, however, quite a lot can be done to correct it by individually feeding the plants with the solutions required for either blue or red flowers as described in the chapter on cultivation.

For example, in most seaside places huge bushes of 'Joseph Banks', often 8 feet high and 10 across, solidly massed with immense corymbs, are commonly found. The flower colour is, however, usually a wretched mixture of muddy purplish-pink, faded blue and rose pink all on the same bush. By individual feeding of the bushes this lamentable appearance can easily be vastly improved. If some of the plants are fed with ¼ oz. of lime and ¼ oz. of iron sulphate per (separate) gallon of water and others are fed with ¼ oz. of aluminium sulphate and ¼ oz. iron sulphate per gallon of water (making one solution) a delightful effect of pure rose-pink and pure pale blue bushes, interplanted, will result.

In the wilder and more naturalistic parts of the garden the Hortensias, like the Hybrid Tea Roses and the hardy hybrid garden Rhododendrons, look out of place. Here the Lacecaps with their half-wild grace and beautifully shaped flowers are supreme. In July, for example, there is surely no flowering shrub more exquisitely beautiful in half shade, than *H. acuminata* var. 'Bluebird'. In August 'Bluewave' follows with its stouter corymbs of less pure colouring but in greater expanse. Aspect is very important, depending upon which of the various ancestral species the hybrid favours in this respect. Thus, in full sun 'Lanarth White' and 'Whitewave' will revel, whereas 'Veitchii', 'Macrosepala' and 'Bluebird' are stunted and miserable.

Too often the woodland garden, a delight in spring, becomes utterly gloomy and dull in summer. This is quite unnecessary. There need be no falling-off in the beauty and quantity of the flowers that it may grow, right up to the time when the autumn tints of the leaves provide the finale. *Azalea macrantha* (*Rhododendron indicum* var. *macranthum*), Azalea 'Daimio' (*R. Kœmpferi* var. 'Daimio'), *Cornus Kousa*, *Rosa Moyesii* and the Genistas bridge the short gap in late June. From July onwards the Hydrangeas of the Lacecap section can provide masses of beautiful flowers at their best in precisely those moist and acid-soiled places where Rhododendrons flourish.

Failure has often attended such plantings in the past for various reasons. In the first place, thoughtful gardeners seeing the seas of pale blue provided by great plantations of 'Joseph Banks' among the trees shading the great Cornish woodland gardens have imported this variety. The result has been disappointing as 'Joseph Banks' will only flourish under trees in the sea-winds and these perpetually blow across our southwestern peninsula. Others have planted the more tender Hortensia varieties in shady positions and slugs and frosts have crippled the display. Others, myself included, have planted Hydrangeas on slopes so steep that the available minerals had been leached away by winter rains and thus the chlorotic and unthrifty plants have made no headway. In other cases failure to protect the plants from slugs during the period when the young growth is soft has been the cause of disaster.

When installing the Lacecaps in sandy or peaty woodland, the soil must be enriched by the addition of some fairly heavy but acid loam, a perforated zinc slug-guard should be placed round the young plant, slug poison should be put down, and the varieties and species should be given the aspect required. If these precautions are taken and the young plants are not allowed to get dry when making their growth it will be found that these Hydrangeas soon make fine bushes and thereafter need little more than routine attention.

Seaside gardens deserve special mention. I have toured our coasts fairly extensively, observing the Hydrangeas grown. At present only three varieties appear commonly. Huge splendid bushes of 'Joseph Banks', small dense bushes of 'Rosea', or 'Mme Chautard' with pale pink flowers and, only occasionally, a lovely tall specimen of 'Mme Emile Mouillère' with the white flowers charmingly flushed with pink and greenish tints. Now, it is precisely in these maritime gardens, where 'Joseph Banks' flowers freely, that the vastly superior hybrids of that parentage should be grown. 'Lumineux', 'Hamburg', 'Mme Legou', 'Monsieur Ghys', 'La France', 'La Marne', 'Paris', 'Altona', etc., are examples. Inland, some of these varieties lose their terminal buds too often to be serviceable outdoors.

To secure the stout growth required to give a profusion of huge corymbs, ample feeding is essential. Otherwise the soil is exhausted just when the final spurt is required to produce a fine mass of flower. Thus it is at the moment when growth starts to be really active that regular feeding must begin. A scattering of complete fertiliser should be given every fortnight and a regular watering every evening in dry weather with weak liquid manure to which the prescribed dose of aluminum sulphate should be added. To facilitate the liquid feeding, a circle of flower pots may be sunk in the ground around the perimeter of each plant. Besides being quickly and easily filled without waste, the pots promote more regular absorption.

As it happens, much of our south coastal strip has a limy soil and here this otherwise disadvantageous position may be turned to advantage to grow the vivid pure reds. Such varieties as 'Heinrich Seidel', 'Princess Beatrix', 'Ami Pasquier', 'Westfalen', 'Président Doumer' and 'Vulcain', especially when they can be given shade from the midday sun, will show a richness of colouring difficult to attain in acid soils.

In modern Close Boskage flowering-shrub beds care is taken in planning so that, whilst 60 per cent of the plants are flowering evergreens and thus weed-deterrent and winter-beautiful, the flower display is equally colourful every week from May to October. To achieve this the 30 per cent of deciduous shrubs permissible must be Hydrangeas – nothing else has the flower-power. There is no room for passengers. As the average effective flowering shrub is 3 feet across the species should be singly interspersed; each gives enough of that particular flower effect in the place at that time. For details see *The Flowering Shrub Garden* by the writer, Farall Publications, Blackdown, Haslemere.

The blue Hydrangeas can easily be grown in terracotta vases or tubs filled with acid soil in such chalky-soiled districts. With all these beautiful plants the private garden may easily put to shame the gaudy little square beds of municipal pelargoniums and bring new beauty to our seaside resorts.

As a plant to keep as a 'pet', so to speak, in a tiny garden, nothing beats a Hydrangea. Its lively personality, its immediate reaction to anything you may do for it and the fascinating different flower colourings, obtained in response to the grower's skill in providing the required soil conditions, make it supremely interesting. I suggest 'Altona'.

Another most satisfactory feature is the way in which a Hydrangea bush, provided it is of a good variety and well cared for, improves steadily all the time. As the roots get down into the moister subsoil, it requires less attention. Every year it is bigger, has more corymbs and these are of a purer blue or a more vivid crimson. Again, the easy propagation of the plant encourages a generosity that is often the first rung on the ladder of friendship. Matters do not end there, for nine times out of ten the recipient plants the young cutting of the azure beauty in lime-polluted soil, and its flowers come a dismal pink. He will lament this misfortune, and then comes the cultivation of real friendship with the help given in the proper cultivation of the plant, and by the time the young offshoot rivals its parent in colour, giver and receiver are already old friends.

In the small personally tended gardens of to-day the Evergreen Azaleas, the Roses and the Hydrangeas are the really important plants. All are the result of centuries of selective breeding and, as a result, bear flowers of such size and brilliance, in such profusion and over so long a period that other kinds are merely secondary.

In the modern garden the hideous 'rosebed' in which the greatest ugliness possible was attained with these beautiful plants by isolating them in square beds of bare earth, gives way to informal associations in which the Roses take their proper places among these other flowering shrubs of similar size and remarkably similar tastes as to soil and aspect. In spring the Evergreen Azalea colourings provide the theme amid the reddish tints of young Rose leaves and the vivid almond green of Hydrangea foliage. In midsummer the Rose is queen until the Hydrangeas take over and remain unchallenged until late autumn. Even in winter their persisting flower-heads decorate the garden attractively against the background provided by the evergreens.

In 1950 a tour of the Midlands and up to Edinburgh revealed that the Hortensias grow well in all counties except in frost-hole areas and the colder parts of Midlothian, Peebles and Selkirk. The importance of selecting free-flowering varieties was, however, very evident, and, in some instances, rank, flowerless bushes showed the evil results of autumn pruning or winter damage without thinning in the following spring.

Chapter 10

Hydrangeas for tubs and vases and as cut flowers

I know of no shrubs that are more effective when grown in fine terracotta vases or even in wooden tubs on a terrace or beside a doorway than the Hortensias. Furthermore, the plants themselves respond so dramatically in their flower colouring to the exact degree of acidity of the easily controlled soil conditions in the containers, that they provide unusual interest in their culture by this method.

No matter how unfavourable the soil of the garden may be, the purest blues, the most vivid purples and the richest red flower colours are easily attained by the specimens grown in the exactly calculated soils thus available.

We have shown how the fundamental pigment colouring of the Hortensia flowers is, with only a few exceptions, simply a crimson pigment and its paleness of tint or its depth and richness of shade depends upon the quantity of pigment in the flowers – varying with the variety. Thus, in a neutral turf-loam we may have the warm pale peach pink which results from a limited number of dots of crimson hue on a white ground – such a tint, in fact, as we see in the *Horticultural Colour Chart* as the palest tint of Crimson (22/3), attained precisely by this means. If we select a deeper coloured variety – say 'Westfalen' – we shall find that, as there are more dots of the same pigment, the flowers will match the full hue of Crimson – *H.C.C.* 22. So much for depth of colour depending on variety.

Now we come to the action of the aluminum, available in acid soil and drawn up by the plant, on the crimson pigment. Here again the readiness to absorb the aluminum (or possibly the quality of vulnerability of the pigment) varies with the variety. Those carrying the colour genes of *H. japonica* resist the blue, those carrying the colour genes of *H. acuminata* are readily affected. Taking a variety such as 'Maréchal Foch', of average readiness to give blue flowers under suitably acid soil conditions, we shall find that all shades from crimson-pink through purple to violet and finally deep blue are attained in exact accordance with rising acidity. Here another factor must be mentioned and that is that blueness builds up. Thus a fine purple will probably gradually build up to pure deep blue in a few seasons.

This correction and adjustment to the desired colouring is a fascinating business. Just now I mentioned a few exceptions to the crimson pigment. Such an exception is the variety 'Vulcain'. Here we have another colour element added; one of great interest as its development may lead us to the production of yellow-flowered Hydrangeas.* This colour element is the range of the xanthophylls – yellow or green leaf colourings derived from the phyllody – or leafy character – of the sepals. With 'Vulcain' we have enough of the yellow tinge to provide an orange-red flower colour streaked with green in neutral soil.

In terrace vases, tubs, or large pots, then, we may, by selection of the best variety for the purpose and then by adjusting the soil in which the plant is grown, achieve Hydrangea flowers of orange-red, crimson, flesh pink, deep pink, lilac, purple, violet, deep blue or pale blue. Thus a collection of Hydrangeas in tubs filled with suitable soils can provide, with little trouble, a range of contrasting flower colours that could only be obtained in the open ground by repeated soil treatments for each individual plant. A position that receives shade for the greater part of the day is desirable. With small varieties no repotting is needed.

Now as to the best containers, large terracotta ornamental pots of many charming old designs – Neapolitan, Tudor or vine-leaved – are undoubtedly the most beautiful and are also eminently practical, but plain pots, preferably at least 15 inches across at the mouth, or even wooden tubs are quite satisfactory. There should be a drainage hole an inch wide for every 6

* See Chapter 5.

inches of bottom space and it is best to cover these, inside, with a couple of inches of the usual crocks.

After the drainage holes are thus protected, a layer of a few inches of well-rotted leaves is next put in to assist in conserving moisture and then the soil mixture, composted to the required acidity, may be added.

The John Innes compost, without the lime unless red flowers are required, is quite satisfactory, but lime-free, fairly heavy, fibrous turf-loam from the top spit of old pasture has remarkable qualities for growing Hydrangeas. In my experience better growth is made in such soil than in any other and I would consider the trouble of gathering a few sackfuls of 'moleheap' earth from the fields very well rewarded. Too black and 'woodsy' a soil does not grow Hydrangeas well, no more does a soil that has received much farmyard dung. Chlorosis (evidenced by sickly yellow leaves) results. Why this should be I do not know, but French growers have the same trouble.

Warning should be given against the yellow subsoil all too often sold as 'loam'. Technically, loam is simply a soil in which both sand and clay particles are found. Infertile loams from too great a depth are most unsatisfactory for Hydrangeas. A fibrous turf-loam, on the other hand, with a very little rotted black leaf-mould or peat to increase the moisture-holding capacity of the mixture, and some sharp *coarse* sand to assist aeration and drainage, are ideal ingredients for the compost.

When filling the containers care should be taken to arrange the soil level so low that there is a good 2 inches depth left free, below the rim, so as to enable watering to be done quickly and conveniently. There is nothing more annoying than to have to wait about, giving a trickle at a time, because the pots are so full of soil that the required dose of liquid cannot be given all at once. At the same time it should be mentioned that Hydrangeas do best when kept on the dry side. Continual flooding is often fatal.

When dry a good soaking is required, and it is, therefore, worth while to have either a hose or a couple of cans and a tap arranged nearby. Rainwater is better than mains water.

Hydrangeas will grow vigorously in very *acid* soils, but they are singularly intolerant of *sour* soil. The latter condition may be described as a soil which is waterlogged or from one cause or another has its particles so closely set together that soil atmosphere is driven out. Therefore, soil organisms cannot exist and the roots are suffocated for lack of air. The plant wilts and droops its flaccid leaves, growth stops and immediate careful repotting with more sand and quicker drainage is necessary, if the plant is not to be lost altogether. Even plants in the open ground may suffer from this condition if over-watered and the remedy, there, is to loosen up the surface soil carefully with a small fork or to push a smoothly pointed stick gently into the soil. It is to assist in avoiding sourness that it is recommended that plenty of coarse sand should be added to the compost, and the addition of a surface mulch of bracken meal will also be found to be an excellent added safeguard.

When large pot-plants from shaded greenhouses or frames are first put out, they should be stood in a calm, shady place for some days before being exposed to full sun and wind. Otherwise the foliage will turn either red or purple (depending on whether the plant is grown red or blue flowered) and the plant will be checked in growth.

Partly forced plants can be put out in tubs after hardening off in late May in most districts. But they do not often attain the rich flower colourings achieved by real open-air plants, though, of course, their flowering is much earlier.

In cold districts the Hydrangeas in tubs may be wintered in a 'coldhouse', or a mere pit covered with a frame light. Every opportunity should be taken to ventilate their winter quarters and even to stand them outside in mild weather. Even so a watch must be kept for aphis, mildew, damping off and all the other ills which attack Hydrangeas so readily under glass.

In the milder districts they may be wintered in a lath-house or even left outside, set fairly closely together with straw or bracken packed between the containers to prevent frost affecting the roots. As soon as weather conditions appear favourable in spring, depending on the local climate, the tubs may be set out in their summer quarters. Dead branches are removed and light watering can commence if the plants appear very dry.

It is important to keep the plants cool enough in autumn to make them drop their leaves naturally – even at the expense of a little frosting. Otherwise the dead leaves have to be cut off and the plants, not having achieved a really dormant period, do not either force or grow away freely at the natural time.

The lath-house is one of the newer developments in horti-culture, born of the new delight many gardeners have found in the culture of such beautiful shade-loving plants as Camel-lias, Rhododendrons and Hydrangeas. Although the structure is not frost-proof, it keeps off the worst effects of winter blizzards and does much to mitigate to mildness the deadly late spring and early autumn frosts. In summer the lath-house provides a cool retreat of lightly sun-dappled shade.

A utilitarian model may easily be made with a lean-to framework of rough poles connected by stout fencing-wire and this may be covered with hop-netting. In small gardens a more ornamental building may be wanted. For this the posts and rafters should be of squared wood, properly halved or mortised and tenoned as required, and a span roof of low pitch will be found to look more attractive. The panels filling in the spaces between the posts and rafters may be formed of sawn builders' laths fixed about ¾ inch apart. When complete, the woodwork may be both decorated and preserved by a coat of one of the green solutions sold for the purpose.

In the lath-house young Hydrangeas may be grown-on to provide fine specimens for tubs and vases later and established plants that are sun-scorched or otherwise damaged may con-valesce. In winter, established tub and vase plants may safely hibernate.

HYDRANGEAS AS ROOM PLANTS

The Hortensias are singularly healthful and effective as room plants in the dwelling house. Suitably prepared specimens in pots (see p. 180) may be taken from cold frames into the living room in January, February, or March. They will soon come into growth, provided that care is taken to give them plenty of light and a good soaking in a bucket filled with sufficient water to come half up the sides of the pot, whenever they look dry. Actually, by picking out suitable plants, I have had them in bloom every month of the year, but the Christmas flowers are often very pale and greenish in colour. Rich-coloured, dwarf varieties like 'Vulcain' or 'King George' are particularly effective. I find it a good plan to give the plants more soil and larger pots than is good practice for other plants. This makes them less sensitive to dryness between waterings and gives them more scope for their tremendous powers of growth. Most room plants are very sensitive to gas leaks, but quite perceptible leaks of Calor Gas do not appear to bother my Hydrangeas. Sometimes the room atmosphere gets too dry for the flower-heads, but a timely spraying with an atomiser usually restores them provided that the damage has not gone too far.

HYDRANGEAS AS CUT FLOWERS

The Hortensias last remarkably well when cut and placed in water. Indeed, I have often had them remain fresh for a fortnight or more. With the blue varieties, a small pinch of aluminum sulphate and one of iron sulphate added to the water assists the colouring of partly opened flower-heads.

If a flower-head wilts this is probably due to an air-lock in the stem. It may often be put right by cutting off a short length of the stem *under water* with secateurs. With the air-lock thus removed the stem sucks again and the head expands perfectly once more.

In late summer, when the heads have assumed their green, bronze, red or purple autumnal tints, they may also be gathered with advantage. Any prematurely browned florets should be carefully snipped off and the head lightly shaken to dislodge the debris.

Placed in water to ripen off finally, such heads will gradually dry-off and remain decorative for a couple of years at least. The colour gradually fades to a buff tone, but the heads may easily be painted crimson, rose-pink, pale blue or dark blue and the stalks green. Thus we have an 'everlasting' flower very attractive as an inexpensive winter decoration when associated with the tinted Butchers Brooms leaves, Statice, etc., generally used in this way.

At Farall Demonstration Gardens, on the east face of Blackdown Hill in West Sussex, but near Haslemere, Surrey, we have a collection of Hydrangeas which, so far as I can discover, is the most complete in Europe. Interested persons are, at all times, welcome to come and see the plants except on Sunday afternoons which are reserved for the playing of the game of Boules, or Pétanque.★

The following species are to be seen:

Hydrangea acuminata
anomala
arborescens
aspera
heteromalla forma *Bretschneideri*
aspera forma *glabripes*
heteromalla
integerrima
involucrata var. *hortensis*
japonica
japonica macrosepala
robusta forma *longipes*
maritima

★ See *The Game of Boules* (*Pétanque and Jeu Provençal*), Farall Publications 1973, by the writer. (Price £1.50 plus p. & p.)

paniculata vars.
petiolaris
quercifolia
aspera subsp. *Sargentiana*
scandens
serrata vars.
aspera subsp. *strigosa macrophylla*
Thunbergii
aspera forma *villosa*

There are also, of course, all the most successful outdoor-growing cultivars and hybrids.

The following species are badly wanted for the collection:
H. Oerstedii
H. scandens subsp. *chinensis* forma *Lobbii*
H. heteromalla forma *Mandarinorum*
H. macrophylla forma *stylosa*
H. aspera forma *Rosthornii*
H. aspera forma *Kawakamii*

General Index

Numbers in italics refer to illustrations

Index of botanical names

Since varieties of garden-bred hybrid Hortensias are listed in alphabetical order on pages 89–121, these varieties have only been included in the Index when referred to elsewhere in text as well.

Numbers in italic refer to illustrations.